A Life Science Lexicon

A Life Science Lexicon

William N. Marchuk
Red Deer College

Wm. C. Brown Publishers

Book Team

Editor *Kevin Kane*
Senior Developmental Editor *Carol Mills*
Production Coordinator *Jayne L. Klein*

WCB

Wm. C. Brown Publishers

President *G. Franklin Lewis*
Vice President, Publisher *George Wm. Bergquist*
Vice President, Publisher *Thomas E. Doran*
Vice President, Operations and Production *Beverly Kolz*
National Sales Manager *Virginia S. Moffat*
Advertising Manager *Ann M. Knepper*
Marketing Manager *Craig S. Marty*
Managing Editor, Production *Colleen A. Yonda*
Production Editorial Manager *Julie A. Kennedy*
Production Editorial Manager *Ann Fuerste*
Publishing Services Manager *Karen J. Slaght*
Manager of Visuals and Design *Faye M. Schilling*

WCB Group

President and Chief Executive Officer *Mark C. Falb*
Chairman of the Board *Wm. C. Brown*

Cover design and interior design by John R. Rokusek

Library of Congress Catalog Card Number: 90–85521

ISBN 0–697–12133–X

Printed in the United States of America by Wm. C. Brown Publishers, 2460 Kerper Boulevard, Dubuque, IA 52001

10 9 8 7 6 5 4

To Bonnie

Contents

Preface

Learning and, more important, understanding a new language can certainly be a challenging task for many students. Learning the language of the life sciences* is no exception, especially for the beginner. It is precisely for this reason that I have written this lexicon. A lexicon is a glossary or compilation of words related to a specific subject area, in this case the life sciences. No previous knowledge of any life science is needed or presumed, and any first-year university/college student should be able to read and use this book with ease and confidence.

Over the years, several experiences have indicated to me a need for such a lexicon. First, many of my students have asked and continue to ask many of the same questions: "What does that word mean? How can you expect me to know what a word that long means? What good will it do if I learn all this technological terminology?" Second, a reference book of this kind is long overdue. To my knowledge there is not one usable, up-to-date reference that a student can carry to a lecture or laboratory situation and use effectively and meaningfully. Either the current reference weighs five kilograms, is too specific to one branch of the life sciences (e.g., Medicine), too cursory (e.g., Introductory Biology), or is encumbered with outdated usages and irrelevant grammar. A third source of evidence for the need for such a lexicon comes from the students who have used excerpts from some of my preliminary manuscripts as lecture/laboratory aids. Many students have indicated that an expanded version of these manuscripts would be most useful.

As I have said, this book has been written to address the needs of many different types of students. Not only will first-year university/college biology students find this book useful, but students in related disciplines such as the allied health fields (e.g., medical technologists, medical-technical writers, journalists, hospital personnel, insurance examiners, nursing and premedical students) and

***The term life sciences is used here to include all the traditional disciplines, such as botany, ecology, genetics, microbiology, zoology (and their subdisciplines) and many of the applied fields, such as agriculture, biotechnology, and medicine.**

laypeople interested in learning and understanding contemporary scientific and technological terminology and using it effectively. We are constantly subjected to various forms of electronic and print media informing us of health, environmental, and technical concerns that we are to make intelligent decisions about. Understanding the language of science and technology would help us with these decisions. Similarly, journalists may find such a lexicon useful in using scientific terminology in the correct and most effective context.

Over the years many students have also asked, "Why Latin and Greek? Why not use common everyday English? Wouldn't it be easier?" My answer is that the western languages arose out of the ancient Roman and Greek languages. Further, very valid, reasons for using Latin and Greek root words include the following:

1. Latin and Greek root words are exact. Vernacular (everyday) language can be confusing. The same word may have several totally different meanings depending on geographical location. Similarly, meanings may change over the years. For example, the word *hacker* can mean

    - the Eastern chipmunk (*Tamias striatus*).
    - one who strikes an opponent's arm in basketball.
    - one who kicks an opponent's shin in rugby.
    - one who breaks up clods of earth in England.
    - one who plays golf poorly.
    - one who uses a computer.
    - one who dissects poorly.

2. Many people erroneously lump quite different animals under the name *bear.*

    - the grizzly bear: *Ursus* (Latin for bear) *horribilis* (Latin for terrible)
    - the koala bear: *Phascolarcts* (Greek for pouched bear) *cinereus* (Latin for ashen-colored)
    - the panda bear: *Ailuropoda* (Greek for cat-like feet) *melanoleuca* (Greek for black and white)
    - the water bear: *Echiniscus* (Latin for little spines) *arctomys* (Greek for mysterious bear)

    Superficially, the animals people describe as bears may resemble one another, but they are not all bears and the scientific names tell you so.

    On the other hand, the cougar, puma, panther, mountain cat, and mountain lion are all the same animal, *Felis concolor,* which is Latin for a cat with color. Clearly, the scientific language has a "universal" application.

3. Vernacular names simply may not exist for some things. For example, endoplasmic reticulum does not go by any other name.

4. Scientific terminology causes few emotional or embarrassing situations. For example, if a biology instructor used the vernacular slang for certain body organs when discussing sexual reproduction, he or she would be dismissed for offensive conduct. If, however, the same organs were described using Latin or Greek, few people would even blush.

A final and most encouraging source of evidence for the need of this lexicon comes from many of my colleagues, including biologists, technologists, English instructors, student counselors, and science editors who have either reviewed my proposal or have commented on the idea. All have indicated that writing this lexicon is a sound and worthwhile endeavor.

The most important and most unique features of this lexicon are:

- its broadness. This lexicon deals with all major disciplines in the life sciences and allied health fields at an introductory level, so it will be applicable to many types of students.
- its emphasis on word construction and usage. A student will be able to look up a particular word root, prefix, or suffix, determine its meaning, and see it used in a contemporary context.
- its size and portability. A student will be able to carry the lexicon to a lecture or lab situation and use it easily on a daily basis.
- its currentness and practicality. The examples and definitions used will be up-to-date and, wherever possible, the latest biotechnical terminology and terms relevant to current trends and concerns will be included.

It is not the intention of this lexicon to impart a thorough knowledge of Latin and Greek grammar. It is written from the standpoint of English. In most cases the root words are given (e.g., -nucle-, -zo-) rather than the actual Latin and Greek words (e.g., *nucis, zoon*) and the definitions given may not be those of the Latin or Greek but rather those that have come to be used figuratively and are of English derivation. For example, cancer in Latin means *crab;* however, the medical derivation arose from the resemblance of the swollen blood vessels surrounding malignant tumors to limbs of a crab. Similarly, suffixes are presented as they are used today rather than according to their original Latin and Greek derivation.

Any comments regarding omissions, errors, or suggestions in this edition will be greatly appreciated. Hopefully many of these will be incorporated into future editions or reprints.

I wish to thank the following reviewers of this edition: Jay M. Templin, Widener University; Rich Blazier, Parkland College; Barbara Yohai Pleasants, Iowa State University; Moe Pushak, Red Deer College; and Anna Okkerse, University of Alberta. I would also like to thank my friend and mentor, Alina Walther.

To all of you new life sciences students—Good luck, work hard, but above all have fun and thank you for choosing this book.

Will Marchuk B.Sc. M.Sc.

1

Introduction

Word Construction

The words used in the life sciences are often long, complicated, and somewhat intimidating. Almost all of these words can, however, be broken down into four basic elements. These elements include **root words** (sometimes called word stems or bases) that may be located anywhere within a particular word. These roots were originally true Greek or Latin words but through time have lost the characteristic endings. A unique feature of this lexicon is that these somewhat meaningless endings, when used in the English language, are removed from all words. For example, the root **-bi-** used today comes from the Greek **bios** meaning *life;* the **-os** ending has been dropped. **Prefixes** are common elements usually found at the beginning of a word attached to a root word and are used to modify the meaning of the root. For example, the common prefix **a-** means *without;* thus the term abiotic refers to something not living. **Suffixes** are frequently found attached to root words, appearing at the end of a word, and are also used to modify the meaning of the root. For example, the common suffix **-sis** means *process of;* thus the term photosynthesis refers to a process that uses light to synthesize something. **Combining vowels** (usually o) are simply used to link the root to the suffix or to another base.

With this basic knowledge of word construction, the meaning of even the longest, most intimidating word can be determined. Several examples will help you get started. You have probably heard of deoxyribonucleic acid (often abbreviated DNA) and have a basic idea of what it is. This is how its construction can be analyzed to determine exactly what the term means:

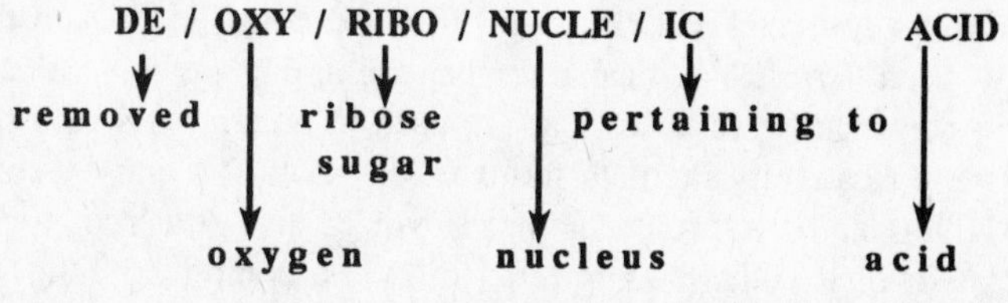

DE / OXY / RIBO / NUCLE / IC ACID

DE → removed
OXY → oxygen
RIBO → ribose sugar
NUCLE → nucleus
IC → pertaining to
ACID → acid

DE is a prefix
OXY is a root
RIBO is a root
NUCLE is a root
IC is a suffix

ACID is a separate word composed of one root meaning sour.

It may seem awkward at first but the meaning of words, when broken down as we have just done, is determined in most cases by reading the word backward from the suffix. Thus the term deoxyribonucleic acid means: **an acidic compound pertaining to the nucleus containing a ribose sugar with one oxygen molecule removed.**

The precise meaning of atherosclerosis can be determined in a similar fashion:

ATHER / O / SCLER / O / SIS

fatty plaque — **combining vowel** — **hard** — **combining vowel** — **process**

ATHER is a root
O is a combining vowel
SCLER is a root
SIS is a suffix

Thus the term atherosclerosis means: **a process associated with hard fatty plaque.** It is a form of arteriosclerosis commonly referred to as "hardening of the arteries."

Remember, the meaning of the roots, prefixes, and suffixes you have just learned will not change, and words with the suffix *-sis* will always refer to a process of some kind.

How to Use this Book

It is obvious that the language of the life sciences can at first glance be somewhat confusing, but if you remember a few simple rules about basic word construction and usage, the time you spend using this book will be time well spent.

Chapter 2 is designed to help you with word construction and the meaning of commonly used terms. For example, the combining vowels (in some instances combining consonants are used) that are used to attach parts of a particular word together are indicated as alternate forms of the root. The entry **-bi-o** indicates that the root is **-bi-** and the usual combining vowel is **o** (the root together with the combining vowel is sometimes referred to as the combining form). The root will appear in words as **-bi-,** or **-bio-,** which means "life."

As previously mentioned, root words may appear virtually anywhere in a word. The root **-bi-o** is an excellent example of this common trend and appears in the following words: **bio**logy, a**bio**genesis, amphi**bi**an, micro**be** (**e** is sometimes used as a combining vowel). To some students the above example may seem to imply that **bio** is sometimes a prefix (**bio**logy), a root (amphi**bi**an), and a suffix (micro**be**). Technically it is a root and to help eliminate any confusion a student may have, Chapter 2 has been written to clearly indicate the true prefixes, roots, and suffixes. Prefixes (and only prefixes, which appear at the beginning of words) are followed by a hyphen: **a-** as in **a**biotic; **ana-** as in **ana**bolic; **hyper-** as in **hyper**tonic. Root words are preceded and followed by a hyphen: **-adapt-** as in **adapt**ation; **-bib-** as in im**bibe**; **-scler-o** as in athero**scler**osis. Suffixes (and only suffixes, which appear at the end of words) are proceeded by a hyphen: **-al** as in abdomin**al**; **-ary** as in coron**ary**; **-logy** as in bio**logy**.

Remember that the meanings of the various root words, prefixes, and suffixes stay the same. Therefore, if you remember the meaning of the various parts of what appears to be a totally foreign word, you should be able to determine the meaning of the entire word.

Chapter 3 provides a list of the common descriptive terms students are most likely to come across in first-year courses. These terms include colors and related qualifying terms, sizes, shapes, textures, direction, position, and quantity. This chapter has a format similar to Chapter 2 in that Latin and Greek roots, the English meaning, and a contemporary example are given for each term.

Chapter 4 provides an alphabetical glossary of terms most often encountered by first-year life sciences students. All the major disciplines are covered including anatomy, botany, cell biology, biochemistry, ecology, microbiology, physiology, and zoology. In addition to a concise definition, each word is broken down into its various components (prefixes, roots, and suffixes) with the meaning of each component given. A unique feature of this chapter is, wherever possible, each word is cross-referenced with its most appropriate synonym and antonym.

Chapter 5 provides a concise description of the major taxonomic categories (kingdom, division or phylum, and selected classes) used in virtually all modern introductory biology textbooks. It is based on the presently accepted "five kingdom system" of classification. In addition to a description of each taxon cited, the meaning of the various components of each taxonomic word is given.

2

Glossary of Common Biological Root Words, Prefixes, and Suffixes

Root / Prefix / Suffix	Usually Means	As In
A		
a-, an-	no, not, without	**a**biotic, **an**aerobic, **A**gnatha, **a**petalous
a-, ab-, abs-	away from, by, to depart	**ab**duct, **ab**normal, **ab**axial, **abs**cess, **ab**ient
abdomen-	belly	**abdomin**al cavity
aberr-	to go astray	**aberr**ant
-able, -ible, -ble, -bul	able to, tending to	adapt**able,** flex**ible**, abl**able**
abras-	rubbed off, scraped off	**abras**ive leaves
abscis-	cut off	**abscis**sion layer
absol-	perfect, complete	**absol**ute alcohol, **absol**ute zero
absorp-	to swallow-	**absorp**tion
abys-	bottomless	**abys**sal
-ac, -aceous	pertaining to, resembling	cardi**ac**, herb**aceous**
-acanth-	thorn, spine	**acanth**aceous, *Polyacantha*
access-	supplementary, extra	**access**ory pigment

Root / Prefix / Suffix	Usually Means	As In
-accre-	grown together, to increase	**accre**tion
-aceae	denoting a taxonomic group (often a family)	Spirochaet**aceae**
-acer-, -acri-	sharp, pointed, bitter	**acri**d, **acer**aceous
-acetabul-	vinegar-cup	**acetabul**ous, **acetabul**um
-acid-	sour, tart	**acid, acid**osis, **acid**ophil
-acin-	grapes in a cluster, berry	**acin**ar cells, **acin**aceous
-acont-	dart	**acont**ia
-acoust-	to listen, to hear	**acoust**ic
-acr-o	summit, top, tip, extreme, sharp	**acro**some, **acro**philous, **acro**miodeltoid muscle
-act-	to do, to drive, to act	**act**ivation
-actin-o	ray, star-like	**actino**morphic, **actino**stele, **actin**
-acu-	sharp, point	**acu**puncture, **acu**leate
-acy	quality, condition of	fall**acy**, accur**acy**
ad-, af-, ag-	toward, to, near, up to, to bring	**ad**ductor, **ad**here, **ad**axial, **af**ferent, **ad**ient, **ag**grade
-ad	toward	cephal**ad**
-adapt-	to fit	**adapt**ation
-addict-	devoted, compelled	**addict**ion
-ade	relating or pertaining to	dec**ade**
-adel-o	obscure	**adelo**morphic
-aden-o	gland	**aden**oids, **adeno**carcinoma, **aden**iform
-adher-	to stick to	**adher**ent
-adip-o	fat	**adipo**se tissue
adn-	to grow on or to	**adn**ation
-adunc-	hooked, bent inward	**adunc**ate
-adventit-	not belonging to, foreign	**adventit**ious roots

Root / Prefix / Suffix	Usually Means	As In
-aer-	air, gas	**aer**ial, **aer**obic, **aer**ate
-aesthe-	see -esthe-	
af-	see ad-	
ag-	see ad-	
-ag-	to do, drive	**ag**ile
-age	collection of, aggregate of	foli**age,** link**age,** cleav**age**
-agglom-	crowded together	**agglom**erate
-agglutin-	glued together	**agglutin**ation
-agoge, -agogue	induce to flow, expelling	lacto**gogue**
-agon-	to fight, struggle against	**agon**y, ant**agon**ist
-agra	painful seizure	arthr**agra**
-agri-	field	**agri**culture
-al, -ial, -eal	pertaining to, like, having the character of	biologic**al,** bacter**ial,** pin**eal**
al-, a	wing	**al**iferous, **al**iform
-alb-, -albin-o	white	**alb**umin, **alb**ication, **albin**o
-ales	denoting a taxonomic group (often a plant order)	Aster**ales**
-aleuron-	flour	**aleuron**ic, **aleuron**e layer
-alg-o	pain	an**alg**esic
-alg-a	seaweed	**alg**ae, **alg**in, **alg**ologist
-alg-e	sensitive to pain	**alg**esis
alien	belonging to another	**alien**
aliment-	food	**aliment**ary canal
-alkal-o	basic, alkaline	**alkal**ine, **alkal**osis
all-o	other, different	**all**opatric speciation
-allant-o	sausage	**allant**oid, **allant**ois, **allant**oic artery
-allel-	of one another	**allel**e

Root / Prefix / Suffix	Usually Means	As In
alpha	alpha (A, α), first Greek letter	the first event in a series or sequence, such as α-hemolysis
-alve-	small, cavity, hollow	**alve**olus
-altric-	nourisher	**altric**es
-am-	to love	**am**orous
-amath-o	sandy soil	**amath**ophile
ambi-	both, around, circulating	**ambi**dextrous, **ambi**ent
-ambly-o	dull	**ambly**opia
-ambul-	walk	**ambul**atory, **ambul**ate
ameb-	change, alternation	**ameb**oid
-amel-	enamel	**amel**oblast
ament-	a strap	**ament**iform
-amine	having a chemical origin, resonous gum	hist**amine,** vit**amin**
-ammon-	the temple of Ammos, where ammonium salts were first prepared from dung	**ammon**ia, **amin**o acid
-amni-	sac, fetal membrane, lamb	**amni**otic fluid, **amni**ocentesis
-amoeb-	change	**amoeb**a, **amoeb**oid
amphi-	on both sides of, both, double	**amphi**bious, **amphi**pathic, **amphi**phloic
-ampli-	to increase	gene **ampli**fication
-ampull-	flask	**ampull**a, **ampull**oid
amyl-o	starch	**amyl**ose, **amyl**oplast
an-	see a-	**an**aerobic
an-	anus	**an**al canal
-an	see -ane	protozo**an**
ana-	up, up against, back, anew	**ana**phase, **ana**bolic
-analog-	ratio, proportionate	**analog**ues, **analog**ous

Root / Prefix / Suffix	Usually Means	As In
-anastomos-	a coming together of two seas	**anastomos**is
-anat-o	cutting, to cut up	**anat**omy
-anc-	elbow, bend of the arm	**anc**oneus muscle
-ance, -ancy	quality, state, condition of	inf**ancy**, dorm**ancy**
-andr-o	male, an old man	**andro**ecium, **andr**ogen
-ane, -an	one connected with, pertaining to	amphibi**an**, hum**ane**
-anemo-	wind	**anemo**tropism, ***Anemone***
-angi-	to contain, vessel, case, enclosed	**angi**tis, gamet**angi**um, **angi**osperm
-angul-	a corner	**angul**ate, **angul**ar
angusti-	narrow	**angusti**form
-anim-	breath, life, spirit, animal	**anim**al
-anis-o	unequal, dissimilar	**aniso**gamy, **anis**form, **aniso**phylly
-ankyl-o	stiff, bent, fusion of parts	**ankylo**sis
-annu-, -ennial	year	**annu**al, bi**ennial**
annul, -annel-	ring	**annul**ate, **Annel**ida
-ans-	loop, handle	**ans**a
-ant, -ent	person who, that which	inhabit**ant**, biological ag**ent**
ante-, anter-	before, prior to	**ante**brachium, **anter**ior
antenna	sailyard	**antenna**
anth-	a flower	**Anth**ophyta, **anth**er, **anth**eridium
-anthrop-o	man, human being	**anthropo**logy, **anthropo**morphic
anti-, ant-	against, opposite	**anti**gen, **anti**body, **anti**septic, **anti**toxin
-antr-	cavity, sinus, cave	**antr**um
-aort-	to raise up (from aeirein)	**aort**ic arch

Root / Prefix / Suffix	Usually Means	As In
ap-o	away, off, away from	**ap**ophysis, **ap**oplast, **ap**heliotropism
-aphan-	obscure	**aphan**ite
-aphrod-	Aphrodite, goddess of love and beauty	**aphrod**isiac, herm**aphrod**itic
-apic-	tip, end, apex, summit	**apic**al meristem, **api**culate
-aplan-o	non-motile, not wandering	**aplano**spore, **aplano**gametes
-append-	hang to, an addition	**append**ectomy, **append**icular
-apsid	arch, a loop	an**apsid, diapsid**
-apt-, -aph-	to touch, to fasten	syn**apt**ic junction
aqua-	water	**aqu**eous solution, **aqua**rium
-ar	pertaining to, like	sol**ar,** son**ar,** pol**ar**
-arachn-	spider, web	**Arachn**id, **arachn**oid
-arbu-, -arbor-	tree-like	**arbor**eous, **arbor**etum
-arc-, -arcu-	bow, arch, bent	**arc**iform, **arcu**ate
arch-, arche-	primitive, first, beginning	**arch**enteron, **arche**gonium, men**arche**
areol-	small space	**areol**ar
-argent-	silver	**argent**ous
-arium, -arion	a place for	terr**arium,** aqu**arium**
-arom-	spicy, fragrant	**arom**atic
-arteri-	artery	**arteri**osclerosis
-arthr-o, -articul-	joint	**arthr**itis, **arthr**opid, **arthro**scopic surgery, **articul**ation
-ary, -ory	a place for, apparatus	avi**ary,** laborat**ory**
-ary	pertaining to, belonging to	involunt**ary** muscle
-aryten-	ladle, pitcher	**aryten**oid cartilage
-asc-o	bag, sac, cup	**asc**us, **asco**spore

Root / Prefix / Suffix	Usually Means	As In
-ase	enzyme	dehydrogen**ase**, sucr**ase**
assimil-	to bring into conformity	**assimil**ation
-asper-	rough	**asper**ous, **asper**ate
aspid-o	shield, broad	**aspid**istra, **aspid**oform
-asthen-	weakness	my**asthen**ia
-astr-, -aster-	star	**aster, astr**onomy, **aster**oid, **astr**ocyte
-ata	characterized by having	plasmodesm**ata**
-atact-o	out of order, disarray	**atact**ostele
-ate, -ite	to treat, to make, to form, characterized by having	dehydr**ate,** sulf**ate,** suf**ite**
-atel-	incomplete, imperfect	**atel**ia
-ather-o	yellow fatty plaque	**athero**sclerosis
-ation	the process or action of	fertiliz**ation**
-atmo-	breath, vapor	**atmo**sphere
-atom-	indivisible, vapor	**atom**ic
-atresia	lacking an opening	proct**atresia**
-atri-	entrance hall, chamber, black	**atri**um, **atri**al
-atten-	to thin out	**atten**uate
-aud-, -audit-	to hear	**audit**ory, **aud**ic nerve
-aug-	to increase	**aug**mentation
aur-	ear	**aur**icle, **aur**iform
auricul-	external ear	**auricul**ar appendage
-austr-	south	**austr**al
-aut-o	self-acting, same	**auto**troph, **auto**matic, **aut**ecology
-aux-o	to increase, to grow	**auxo**spore, **aux**in, **auxo**chrome
-av-	bird	**av**ian, **av**is, **av**iary
-ax-, axon	center line, axis, axle	ab**ax**ile, ad**ax**ial, **axon, ax**ial

Root / Prefix / Suffix	Usually Means	As In
-axill-	armpit	**axill**ary
azot-o	nitrogen	*Azotobacter*
-azyg-	unpaired, unmarried	**azyg**ous twins

B

Root / Prefix / Suffix	Usually Means	As In
-ba-	to step, to walk, to go	hypno**ba**tia
-bacc-	berry, pearl	**bacc**ate, **bacc**iform
-bacill-	a little stick	**bacill**us
-bacter-	small rod	**bacter**ia
-balan-	acorn, penis	**balan**ic
-ball-	to throw, to put	**ball**istics, **ball**istospores
-bar-	pressure	**bar, bar**ometer
-barb-	beard	**barb, barb**ate
-bas-i, -basid-	base, pedestal	**bas**ipetal, **basid**iospore, **basid**ium, **bas**al
-bathy-	depth, deep	**bathy**metry, **bath**osphere
-be	see -bi-	
bene-	well, good	**bene**ficial
benign	kindly	**benign** tumor
-benth-o	depth, bottom	**benth**os, **benth**ic, **benth**ophyte
-bet-	beta (Β, β), second Greek letter	the second in a series or sequence, (β–hemolysis)
-bi-o, -be	life	**bi**ology, anti**bi**otic, amphi**bi**an, micro**be,** aero**be**
bi-, bin-o	twice, double, two	**bin**omial, **bi**ceps, **bin**ary fission, **bin**ocular vision
-bib-	drink	im**bib**ition
-bil-i	bile, gall	**bil**irubin

Root / Prefix / Suffix	Usually Means	As In
-blast-o	sprout, germ, shoot, bud	**blast**ula, **blast**oderm, fibro**blast**
-ble, -bul	see -able	
-blenn-o	mucus	**blenn**orrhea, **blenn**oid
-blep-	to see	a**blep**sia
-bol-	to throw, put	meta**bol**ism, ana**bol**ic
-bol-	lump	**bol**us
-bore-	north	**bore**al forest, aurora **bore**alis
-botan-	grass, fodder, an herb	**botan**y
-botry-	bunch, cluster of grapes	**botry**oidal, ***Botrychium***
-botul-	sausage	**botul**ism, **botul**iform
-bov-	cow	**bov**ine
-brachi-, -brachial-	arm	**brachi**opod, **brachi**ation, **brachial**is muscle
-brachy-	short	**brachy**sm, **brachy**pterous
-bract-	thin plate	**bract**eiform, **bract**eate
-brady-	slow	**brady**lexia, **brady**cardia
-branchi-, -branch	gills	**branchi**al arch, **branchi**form, nudi**branch**
-brev-	short	**brev**ity
-brom-	stench	**brom**ide, **brom**ism
-bronch-	windpipe	**bronch**us, **bronch**itis
-bront-	thunder	**bront**osaurus, **bront**esis
-bry-	to grow, to be full of life	em**bry**o
-bry-o	moss	**bry**ophyte, **bry**ology
-bucc-	cheek, mouth	**bucc**al cavity
-buffe-	to blow	**buffe**r
-bul-, -bulum	will, determination	a**bul**ia
-bulim-	hunger	**bulim**ia

Root / Prefix / Suffix	Usually Means	As In
-bull-	bubble, blister	**bull**iform cell, tympanic **bull**a, **bull**ate
-bun-	mound	**bun**odont
-buoy-	to float	**buoy**ancy
-burs-	bag, pouch	**burs**a
-butyr-	butter	**butyr**ic acid

C

Root / Prefix / Suffix	Usually Means	As In
-cac-, -kak-	bad, vile	**cac**hexia
cadaver	a dead body	**cadaver**
-caec-, -cec-	blind end, blind intestine	**caec**um (**cec**um)
-caffe-	coffee	**caffe**ine
-cal-, -calor-	to be warm, heat	**calor**ie, **calor**imeter
-cal-	see intercal-	
-calc-	limestone, calcium	**calc**ification, **calc**areous
-call-o	hard skin	**call**us, corpus **call**osum
-calcan-	heel	**calcan**eus
-caly-	cup, husk, pod, covering	renal **calyx**, **caly**coid
-calypt-, -calyptr-	hidden, a veil	**calyptr**a
-camb-	change, exchange	**camb**ium
-camp-, -campt-	bent, curved, bell-shaped	**camp**aniform, **campt**osaur
-campan-o	bell	**campan**ulate
-campest-	field	**campest**ral
-cancell-	latticed	**cancell**ous bone, **cancell**ate
cancer-	crab, a spreading sore	**cancer**ous
-cani-	dog	**can**ine
cap-, -cept-	to take, to seize	**cap**tivate

Root / Prefix / Suffix	Usually Means	As In
-capi-, -capit-	head	de**capit**ate, **capit**ulum
-capill-	hair	**capill**ary
-caps-	box	en**caps**ulated
-carap-	shell, covering	**carap**ace
-carb-, -carbon-	charcoal, coal	**carbon**iferous, **carbon**ate
-carcin-o	a crab, cancer, spreading sore	**carcin**oma, **carcin**ogen
-card-o	hinge	**card**o
-cardi-o	heart	**cardi**ac, **cardia**, electro**cardi**ogram
-cardinal-	red, principal	**cardinal** vein
-car-, -carn-	meat, flesh	**carn**ivore, **carn**al
-cari-	decay	**cari**on, dental **caries**
-carin-	keel	**carin**ate
-carot-	carrot	**carot**ene
carotid	heavy sleep, to stupify	**carotid** artery
-carp-	fruit	**carp**el, **carp**ellate
-carp-	wrist	**carp**al bones, **carp**us
-cartilag-	gristle	**cartilag**inous
-cary-, -kary-	nucleus, nut, kernel	**cary**opsis, eu**cary**otic (eu**kary**otic), **kary**ogamy
-cas-	to fall	**cas**cade
-cas-	cheese	**cas**ein
cata-	down, against, dissolution	**cata**bolic, **cata**lyst, **cata**genesis
-cau-, -caus-	to burn	**cau**terize, **caus**tic
-caud-o, -cauda-	tail	**caud**al fin, **caud**ate, **caudo**femoralis muscle
-caul-	stem, stalk	**cauli**flower, a**caul**escent, **caul**ine
-cauter-	to burn	**cauter**ize

Root / Prefix / Suffix	Usually Means	As In
-cav-, -cavit-	hollow	**cavit**ation, vena **cava**
-cec-	see -caec-	
-cede-, -cess-	to go, to yield, to follow after	re**cede**, re**cess**ive
-cele, -coele	swelling, hernia	arthro**cele**, gastro**coele**
-celi-	hollow, abdominal cavity	**celi**ac artery
-cell-, -cellul-	a small cell (room)	**cell**, **cellul**ose
-cene	new, recent	Pleisto**cene**
-cente-	to puncture surgically	amnio**cente**sis
-centi-	one hundredth	**centi**metre, **cent**ury
-centri-, -centrum-	center of a circle, middle, a point	**centri**pital, **centri**ole, **centr**omere, **centrum**
-cephal-	head, brain	**cephal**ization, **cephal**ic, dien**cephal**on
-cept, -ceptor	to take	ac**cept**or molecule
-cer-	wax	**cer**aceous
-cerat-o	horn	**cerat**oid, **cerat**otrichia
-cerc-	tail	hetero**cerc**al fin, **cerc**aria
-cerebr-	brain	**cerebr**al, **cerebr**al palsy
Ceres	goddess of grain	**cere**al
-cern-, -cret-	to separate, to secrete	dis**cern**ment, in**cret**ion
-cern-	drooping	**cern**uous, **cern**ate
-cervic-, -cervix-	neck, neck-like	**cervic**al vertebrae
-cess-	see -cede-	
-cet-	whale	**cet**ology
-chaet-	hair	**chaet**iferous, **chaet**ae
-chel-	claw, hoof	**chel**ate, **chel**icera, **chel**iform
-chem-	chemical, drug	**chem**osynthetic, **chem**otaxis

Root / Prefix / Suffix	Usually Means	As In
-chiasm-	cross	**chiasm**ata, optic **chiasm**a
-chil-, -cheil-	lip	**chil**opods
-chir-, -cheir-	hand	**chir**opractic medicine
-chit-	tunic	**chit**in
-chlamyd-	cloak, envelope	*Chlamydia*, a**chlamyd**eous
-chlor-o	green	**chloro**plast, **chlor**ine, **chloro**sis
-choan-o	funnel	**choan**a, **choano**cyte
-chole-	bile, gallbladder	**chole**ra, ductus **chole**dochal, **chole**sterol
-chondr-, -chondri-	cartilage, granule	**chondr**ocyte, **chondri**tis, mito**chondri**a
-chor-, -chori-	membrane, skin	**chor**oid plexus, **chori**on layer
-chord-	cord, string	noto**chord**, **Chord**ata
-chrom-o, chromat-	color	**chromo**plast, **chromat**ophore, **chromo**some
-chron-	time	syn**chron**ize, **chrono**logy
-chrys-o	golden	**Chryso**phyta
-chym-, -enchym-	juice, infusion	**chym**e, par**enchym**a
-cid-, -cide-	to kill, to fall	sui**cide**, geno**cide**, de**cid**uous
-cili-	a small hair, eyelash	**cili**a
-ciner-	ashes	in**ciner**ate, tuber **ciner**eum, **ciner**eous
-cingul-	girdle	**cingul**um
-cipit-	headlong	pre**cipit**ate
-circ-, -circa-, -circum-	ring, round, about, around	**circ**ulate, **circum**polar, **circa**dian
cirr-o	hairlike curls	**cirr**us
-cirrh-o	tawny (orange-yellow)	**cirrh**osis

Root / Prefix / Suffix	Usually Means	As In
cis-	to cut, to kill	incision
-cist-, cistern	a box, chest	**cist, cistern**
-clad-i	branching	**clad**istic, **clad**ophyll, **clad**ogensis
-clast-, -clasia	to break	**clast**ic, arthro**clasia,** osteo**clast**
-claus-	to close	**claus**trophobic
-clav-	club	**clav**ate
-clavic-	key, collarbone	**clavic**le, sub**clav**ian artery
-cle	little	ventri**cle**
-cleav-	to divide	**cleav**age furrow
-clei-, -cleist-	close, shut	**cleist**othecium
-cleithr-	key, bar	**cleithr**um
-clin-	to lean	de**clin**e, in**clin**e
-clitell-	pack saddle	**clitell**um
-clitor-	to enclose	**clitor**is
-cliv-	slope	**cliv**us
-cloac-	sewer	**cloac**a
-clon-, -klon-	twig	**clon**ing, mono**clon**al antibody
-clos-, -clud-	to shut, to close	con**clus**ion, oc**clud**ed
-clys-	to wash out, irrigate	cata**clys**m, colo**clys**is, **clys**ter
-clyp-	shield	**clyp**eate
-cnem-	shin	gastro**cnem**ius muscle
-cnid-	stinging nettle	**cnid**ocyte
co-	together, with	**co**enzyme, **co**valent, **co**evolution
coacerv-	to pile up, amass	**coacerv**ate droplets
-coagul-	drive together, curdle	**coagul**ation, **coagul**ase
-coales-	to grow together, fuse	**coales**cence

Root / Prefix / Suffix	Usually Means	As In
-cocc-i	a berry, seed	**cocc**i, *Staphylococcus*, **cocc**oid
-coch-	land snail	**coch**lea
-coel-o	abdominal cavity, hollow	**coel**om, **coel**enterate, **coel**iac artery, pseudo**coel**om
-coen-	shared in common	**coen**cytic
-cog-	to think, consider	**cog**nitive
-cohes-	to stick together	**cohes**ion
-coit-	sexual conjugation	**coit**us
-col-	colon, large intestine	**col**itis, **col**ectomy
-col-	strainer, sieve-like	**col**iform
-col-	see com-	
-cole-	sheath	**cole**optile
-coll-	neck	**coll**ar, longus **coll**i muscle
-coll-a	glue	**coll**agen, **coll**enchyma, **coll**oidal
colon-	farmer	**colon**y
-colp-	vagina	**colp**iti
com-, con-, col-	with, together, dwell	**com**mensal, **con**genital, deserti**col**ous
-commissur-	connection, point of union	**commissur**e
-commun-	living together	**commun**ity
-comple-	to fill up	**comple**mentation
compo-	to put together	**compo**und eye
-con-	cone	**con**us arteriosus, **con**iferous
-concep-	to conceive	**concep**tacle
-conch-	shell	**conch**iform
-condens-	to press close together	**condens**ation reaction
-condyl-	knob, knuckle	epi**condyl**e
-confer-	crowded	**confer**ted

Root / Prefix / Suffix	Usually Means	As In
-coni-	dust	**coni**diophore
-conju-, conjunctiv-	to join together	**conju**gation, **conjunctiv**um
constipat-	to press together	**constipat**ion
-constrict-	drawn together	**constrict**ion
contra-, counter-	opposite, against	**contra**ceptive, **counter**stain
-converg-	to come together	**converg**ence
convol-	to roll or twist	**convol**uted
-copr-, -copra-	dung, excrement	**copr**ophagia, **copr**odeum
-copul-	a link, bond	**copul**ate
-cor-, -core-	pupil of the eye	**cor**nea
-corac-o	raven, crow, beak-like	**corac**oid
-cord-	heart	**cord**ate
-cori-	skin, leather	**cori**um, **cori**aceous
-corm-	trunk (of a tree)	**corm**, **corm**ophytic
-corn-	horny, keratinized tissue	**corn**ate, **corn**ify, **corn**ea
-coroll-	little crown	**coroll**a
-coron-	crown, something curved	**coron**ary arteries
-corp-, -corpus-	body	**corp**se, **corpus**cle
-correlat-	relationship	**correlat**ion
-corro-	to gnaw	**corro**sive
-cort-, cortex	bark, shell, covering	renal **cortex**, **cort**icate, **cort**ical
-corymb-	a cluster of flowers	**corymb**ate
-cosm-o	world, universe	**cosmo**politan, **cosmo**s
-cost-	rib	**cost**al cartilage
-cotyle-	cup	**cotyle**don
-cox-	hip	**cox**itis, **cox**al
-cra-	to mix, mixing bowl	**cra**teriform
-cran-, -crani-	skull, cranium, brain	**crani**um, **crani**al
-cre-	flesh	ex**cre**ment, pan**cre**as

Root / Prefix / Suffix	Usually Means	As In
-crem-	burn	**crem**ate
-cren-	notched, scalloped	**cren**ation
-crepit-	to creak	**crepit**us
-cresc-	to grow	**cresc**ent
-cret-	to separate, to sift	se**crete**, ex**crete**
-cri-	to separate, to distinguish	**cri**teria, **cri**tical
-cribr-	sieve-like	**cribr**iform plate
-cric-	ring	**cric**oid cartilage
-crin-	hair	**crin**ose
-crine	to secrete, to separate	endo**crine**, exo**crine**
-crist-	crest, ridge, shelf	**crist**a
-crit-	judging	**crit**ical
-cruci-	cross	**cruci**ate, **cruci**form
-crur-	lower part, leg	abductor **crur**is muscle
-crust-	encrusted, shell	**crust**ose, **Crust**acea
cry-o, crym-	ice, snow, cold	**cry**oflora, **cry**obiology
-crypt-	hidden	**crypt**ic coloration, **crypt**ogam
-cten-o	comb	**cten**oid
cub-	cube	**cub**oidal
-cult-	to develop, to care for	agri**cult**ure, aqua**cult**ure
-cumb-	to lie	re**cumb**ent, de**cumb**ent
-cumul-	heaped	**cumul**ative
-cune-	wedge	**cune**ate, **cune**iform bone
-cup-	tub	**cup**ulate
-cur-	care, cure	se**cure**
-cur-, -curr-	to turn, to go, to run	ex**curr**ent, in**curr**ent
-cusp-	point, flap	bi**cusp**id, **cusp**ate
-cuss-	to shake, to strike	concussion

Root / Prefix / Suffix	Usually Means	As In
-cut-, -cuti-	skin	sub**cut**aneous, **cuti**cle, **cut**in
-cyan-	blue	**cyan**obacteria, **cyan**ic
-cycl-o, -cycle	circle, ring, cycle	**cycl**ic phosphoralation, **cycl**oid scales, peri**cycle**
-cye-	to be pregnant, conception	**cye**siology
-cym-, -kym-	wave	**cym**ograph (**kym**ograph)
-cym-	young sprout	**cym**e, **cym**ose
-cyn-	dog	**cyn**ophobia
-cyst-	bladder, bag	blasto**cyst**, pneumato**cyst**
-cyt-o, -cyte	a hollow vessel, cell	**cyt**oplasm, **cyt**ology, lympho**cyte**

D

Root / Prefix / Suffix	Usually Means	As In
-dacry-o	tear	**dacry**orrhea, **dacry**oid
-dactyl-	fingers, toes	a**dactyl**y, **dactyl**oid
de-	to remove, undoing, down, out, from	**de**aminate, **de**hydration, **de**oxyribonucleic acid, **de**composer
de-	completely	**de**siccation
dec-, deca-	ten	**deca**metre, **deca**hydrate
decem-, deci-	tenth	**Decem**ber, **deci**metre
deferens	carry away	vas **deferens**
-degrad-	to break down	**degrad**ation
decidu-	to fall off	**decidu**ous
dehisc-	to split	**dehisc**ent, in**dehisc**ent
deliques-	to become fluid	**deliques**cent
delt-o	delta (Δ,δ), fourth Greek letter	**delt**oid
dem-	people, country	**dem**e, epi**dem**ic, **dem**ographic
dendr-o	a tree	**dendr**ite, **dendr**oid
dent-, dens-	tooth	**dent**in, **dent**ary bone

Root / Prefix / Suffix	Usually Means	As In
-depres-	to keep down	**depress**or muscle, **depress**ant
-derm-, -dermat-	skin	epi**derm**is, **dermat**ologist, endo**derm**is
-desis	binding, fusion	arthro**desis**
-desm-	to tie, to bind, ligament, to bond	syn**desm**ology, **desm**osome, **desm**id
-determin-	to end completely	**determin**ation, in**determin**ate growth
-detrit-	worn off	**detrit**ivore
-deum	road, on the way	uro**deum,** stomo**deum,** procto**deum**
-deut-, -deuter-o	other, second	**deuter**ostome
-dextr-o	right-hand	**dextr**al, **dextr**ose
di-, dich-, dis-	two, twice, double	**di**peptide, **dich**otomous, **di**ploid, **di**oecious, **di**oxide, **di**cotyledon, **dis**ogeny
diabet-	a siphon	**diabet**ic
dia-, dien-	across, between, through, apart	**dia**lysis, **dia**phram, **dia**betes, **dien**cephalon
diastol-	difference, a separation, to expand	**diastole**, **diastol**ic pressure
-dicty-o	a net	**dicty**osome, **dicty**ostele
-didym-	twin, testicle	epi**didym**is
-diem	day	circa**dian**
-diet-	way of life	**diet**ary
diffus-	to pour out, apart	**diffus**ion, **diffuse**
-digest-	dividing, separating out	in**digest**ion
-digit-	finger, toe	**digit**al
-dilat-, -dilatat-	to expand, enlarge	**dilat**ion
din-o	terrible	**dino**saur
dino-	whirling	**dino**flagellate

Root / Prefix / Suffix	Usually Means	As In
dipl-o	double, twofold	**dipl**oid, **dipl**oblastic
-dips-	thirst	**dips**ophobia
dis-, di-	apart, in different directions	**dis**sect, **dis**location
-disk-, -disc-	disc	**disc**oid, **disc** florets
-dissem-	to scatter seeds	**dissem**ination
-dist-	apart, distant, remote	**dist**al
-diverg-	to bend away	**diverg**ent evolution
-divert-	to turn aside	**divert**iculum
-do-	to give	**do**sage
-doc-, -doct-	to teach	**doct**or, in**doct**rination
-dol-	pain	**dol**ar
-dom-	house	**dom**estication
-domin-	ruling	**domin**ant gene
-dont	see -odont-o	
-dorm-	to sleep	**dorm**ant
-dors-, -dorsi-	back	**dors**al, longissimus **dorsi** muscle, **dorsi**ventral
-drepan-	sickle	**drepan**ocyte, **drepan**iform
-drom-	running, course	syn**drom**e, **drom**edary, ana**drom**ous
-dros-	dew	**dros**ophilic, *Drosophila*
-dru-	bump	**dru**se
-du-	double	**du**plicate
-duc-, -duct-	to lead, to draw	ab**duc**ens, **duct**ile, ovi**duct**
-dulc-	sweet	**dul**se
-dum-	bush	**dum**etose
-duoden-	twelve	**duoden**um
-duplic-, -duplex-	double	**duplic**ation
-dur-	hard	in**dur**ation, **dur**a mater
dyn-	power	**dyn**amics, **dyn**ein arm

Root / Prefix / Suffix	Usually Means	As In
-dys-	bad, abnormal, difficult, hard to do, pain	**dys**trophy, **dys**entery

E

Root / Prefix / Suffix	Usually Means	As In
e-, ec-, ef-, eff-, ex-	out of, away from, external to	**ec**topic, **e**paxial, **eff**erent, **ex**hale
-eal, -ial	pertaining to	arbor**eal**, fil**ial**
-eburn-	ivory	**eburn**ous
-ec-o, -eci-, -oec-, -oeci-	household, home, dwelling	**eco**logy, **eco**system, di**oeci**ous
-ecdys-	slipping out, shifting out	**ecdys**is
-echin-o	spiny, hedgehog	**echino**derm, **echin**ate
ecto-	external, outside	**ecto**derm, **ecto**parasitic
-ectomy	cut out, remove	append**ectomy**
-ectop-	displaced, foreign	**ectop**ic pregnancy
-ed-	to eat	**ed**acious
-edaph-	soil	**edaph**ic
-edem-	to swell	**edem**a
ef-	see e-	
-eil-	to roll	**eil**oid
-eis-	see -es-	
-ejacul-	to shoot forth, thrown	**ejacul**ation
-elai-o	olive oil	**elaio**plast
-elasm-	a thin plate, to draw out	**Elasm**obranchia, **elasm**oid
-elat-	driver, tall	**elat**er
-electr-	amber	**electr**on, **electr**olyte
element-	first principle	**element**ary
-ellum	little	flag**ellum**, cereb**ellum**
-em-	blood	hypoglyc**em**ic
em-	see en-	
-emet-	to vomit	hyper**emes**is, **emet**ic

Root / Prefix / Suffix	Usually Means	As In
-embol-	stopper, wedge	**embol**ism
-emia, -haemia	condition of the blood	hypoglyc**emia,** septic**emia**
-emul-	to milk	**emul**sify
en-, em-	in, into, inward	**en**doscopic surgery, **en**zyme, **em**bryo, **en**vironment
-ence, -ency	the condition of	senes**cence**, val**ency**
-encephal-	brain	**encephal**itis
-enchym-	see -chym-	
endo-, ento-	within, inner	**endo**cytosis, **endo**crine, **endo**skeleton, **endo**spore, **ento**cyclic, **ento**parasite
-ene	denoting an open-ended, unsaturated hydrocarbon	xyl**ene,** ethyl**ene**
-energ-	active	**energ**y
-enne-, -ennea-	nine	**ennea**d
-enni-	year (from annu-)	bi**enni**al
-ens-	sword	**ens**iform
-ent	performing the action of, having the quality of	arboresc**ent**
-enter-o	intestine, gut	**enter**on, **entero**nephric, gastro**enter**itis
-ento-	see endo-	
-entom-o	insect	**entomo**logy
-eo-, -eos-	early age, dawn of, rosy	**eo**sin, **eo**ns, **Eo**cene
-eous	nature of, like	ign**eous,** acerac**eous**
-ephemer-	lasting for a day, short-lived	**ephemer**al
epi-	upon, above, on, over	**epi**dermis, **epi**phyte, **epi**condyle
equ-	equal	**equ**atorial plane, **equ**ilateral
-equ-	horse	**equ**istrian

Root / Prefix / Suffix	Usually Means	As In
-er, -ist, -ast	one who, specialist, connected with	geograph**er,** dent**ist,** Mon**era**
-er-, -erot-	love	**erot**ic
-erect-	upright	**erect**ion, pilo**erect**ion
-erg-, ergist-	work	ex**erg**onic reaction, en**ergy**, syn**ergist**
-err-	to wander, to deviate	**err**atic
-ery	a place of business	fish**ery,** nurs**ery**
-erythr-o	red	**erythr**ocyte
-es-, -eis-	inward, into	esoteric
-esc, -esce	to begin, to be somewhat, to grow	convale**sce,** phosphore**sce**
-escent	beginning to be, have, or do	efferv**escent**
esophag-e	gullet	**esophag**us
-esthe-, -aesthe-	to feel, to perceive	**aesthe**tics, **esthe**sis
-esthen-	weakness	my**esthen**ia
-estr-o	mad desire	**estr**ogen
-et, -ette	little, small	flor**et**
-eth-o	custom, habit	**eth**ology
-ethm-o	sieve	**ethm**oid bone
-ethn-o	people, nation	**ethn**obiology
-eti-o	cause	**eti**ology
-etic	pertaining to	antidiur**etic**
-etiol-	pale, whitish, to blanch	**etiol**ation
-eu-	good, true, well, fully developed	**eu**trophic, **eu**karyotic
-eury-, -eurys-	wide	**eury**baric
-evolut-	an unrolling, to unfold	**evolut**ion
ex-o, ec-, ect-o, e-	out of, away from, external to	**ex**ocytosis, **ex**cretion, **e**jaculation, **e**viscerate, **ect**oplasm, **e**caudate
-expuls-	driven out	**expuls**ion

Root / Prefix / Suffix	Usually Means	As In
ext-, extra-	outside, beyond	**extra**embryonic, **ext**ernal, **extra**cellular
extant-	to stand out	**extant**
extinct-	to be extinguished	**extinct**ion
-exu-	to cast off, undress	**exu**viate
-exud-	sweat	**exud**ate

F

Root / Prefix / Suffix	Usually Means	As In
-fab-	bean	**fab**iform, **fab**aceous
-fac-, -fact-	to do, to make, capability	arti**fact, fac**ulative
-faci-	face, surface	inter**face**, **facie**s
-facilit-	easy, easy to do	**facilit**ation, **facilit**ated diffusion
-falc-	sickle	**falc**iform ligament
-fall-, -fals-	to deceive	**fall**acy
fang	grip	**fang**
-farin-	flour	**farin**ose
-farious	arranged in rows	bi**farious**
-fasc-i	band, bundle	**fasc**ia, **fasc**icle
-fauc-	throat	**fauc**es
-faun-	god of the woods	**faun**a
-fav-	honeycomb	**fav**eolate
-febr-	fever	**febr**iphobia
-fec-	excrement, sediment, dregs	**fec**aloid, de**fec**ation, **fec**es
-fect-	to make, to bring about	ef**fect**or
-fecund-	fruitfulness	**fecund**ity
-feli-	cat	**feli**ne, ***Felis***
-fem-	thigh	**fem**or, **fem**oral artery
-fenestr-	window, opening	**fenestr**ate, **fenestra**

Root / Prefix / Suffix	Usually Means	As In
-fer-, -ferre	convey, to bear, to carry	semini**fer**ous, ef**fer**ent artery, **fer**tile, bi**fer**ous, coni**fer**
-ferment-	to boil, leaven, ferment	**ferment, ferment**ation
-ferr-	iron	**ferr**ous, **ferr**uginous
-fertil-	fertile, to produce	**fertil**ization
-ferv-	to bubble, boil	ef**ferv**escent
-fet-	the young in the womb, pregnant	**fet**us, **fet**al
-fibr-	fiber	myo**fibr**il, **fibr**iform
-fibul-	buckle, pin	**fibul**a
-fic-	to make, to render, to cause	nitri**fic**ation, felici**fic**
-fid-	see -fiss-	
-fil-	thread	**fil**amentous, **fil**iform, **fil**arial worms
-fili-	fern	**fili**cales
-filia-	daughter	**filia**l generation
-fim-	dung	**fim**icolus
-fimb-	fringe	**fimb**oid
-fin	end	**fin**al
-fiss-, -fid-	to split, cleft	**fiss**ion, **fiss**ure, multi**fid**us muscle
-fix-	to put in place	nitrogen **fix**ation
-flabell-	fan	**flabell**iform
-flacc-	flabby, soft	**flacc**id
-flagell-	a whip	**flagell**a
-flamm-	flame	in**flamm**ation
-flat-	to blow	**flat**ulence
-flav-	yellow	**flav**onoids, ribo**flav**in
-flex-	to bend	re**flex**, in**flex**ible
-flocc-	tuft of wool	**flocc**ulation, **flocc**ulose

Root / Prefix / Suffix	Usually Means	As In
-flor-, -floresc-	flower, blooming	**flor**al, **flor**et, in**floresc**ence
-flu-, -flux-	to flow, stream	**flu**id, in**flu**enza, in**flux**
-foli-	a leaf	**foli**ose, **foli**age, **foli**ate papillae
-follic-	small bag	**follic**le
-for-, -foram-	to bore, to pierce, an opening	**foram**ina, **foram**ine magnum
-form-	shape, rule	reni**form**, **form**ula
-formic-	an ant	**formic** acid, **formic**ide
-forn-	arch, vault, (also brothel)	**forn**ix, **forn**ication
-foss-	ditch, trench, to dig up	**foss**a, **foss**il
-fov-	a pit, depression	**fov**ea
-fract-	to bend	re**fract**ile
-frag-	to break	**frag**ile, **frag**mentation
-frat-	a brother	**frat**ernal twins
-frig-	cold	**frig**id
-frimb-	fringe	**frimb**riate
-frond-	a leaf, foliage	fern **frond**, **frond**iform
-front-	forehead, front	**front**al lobe
-frug-, -fruct-	fruit	**fruct**ose, **frug**ivorous
-frust-	a piece	**frust**ule
-frutic-	shrub	**frutic**ose
-fuc-	a seaweed	**fuc**oxanthin, ***Fucus***
-fug-	to flee	centra**fug**al, **fug**acious
-fum-	smoke	**fum**igate
-fun-	cord, rope	**fun**icular, **fun**iform
-funct-	to perform	uni**funct**ional
-fund-	bottom, base	gastric **fund**us, in**fund**ibulum
-fung-	a mushroom	**fung**us, **fung**iform papilla
-funicul-, -funi-	string, fiber	**funicul**us

Root / Prefix / Suffix	Usually Means	As In
-furc-	a fork	bi**furc**ate
-fus-i	spindle	**fus**iform
-fus-	to pour	trans**fus**ion
-fusc-	dark, brown	**fusc**in
-fy	to make, to cause	liqui**fy**, putre**fy**

G

Root / Prefix / Suffix	Usually Means	As In
-galax-, -galact-o	milk	**galact**ose, **galax**y
-gale-o	helmet	**gale**ate
-gam-, -gamet-	marriage, union, spouse	mono**gam**y, **gamet**es, **gamet**ophyte
-gangli-	knot, small cyst, swelling	**gangli**a
-gastr-o, -gaster-	stomach, belly of muscle	**gastr**opod, **gastr**ic, **gastro**cnemius muscle
-ge-o, -geo-	earth	**ge**otrophic, bio**ge**ography
-gel-	congealed	**gel**atinous
-gemin-	twin-born	tri**gemin**al nerve
-gemm-	bud	**gemm**ule
-gen-o, gener-	producing, origin, beginning, descent, race, birth	**gen**etics, **gen**ocide, **gen**us, **gener**a, gameto**gen**sis, **gen**otype, hydro**gen**, endo**gen**ous, onto**gen**y, partheno**gen**esis
-gen-	knee	**gen**iculate
geny-, geni-	jaw, chin	**geni**ohyoid muscle
-geo-	see -ge-o	
-ger-, -geront-	old person, old age	**geront**ology
-germ-, -germin-	to sprout, bud	**germin**ation
-gest-a	to carry, to bear	**gest**ation
-geu-	to taste	hyper**geu**sia
-gibb-	humpbacked, bent-over	**gibb**ous
-gingiv-	the gums	**gingiv**itis

Root / Prefix / Suffix	Usually Means	As In
-girdl-	to tie with a belt	**gird**ling
-glabr-o	smooth, without hair, bald	**glabr**ous
-glac-	ice	**glac**ial
-gladi-	sword	**gladi**olus
gland-	an acorn	**gland**iform, **gland**ular
-glauc-o	bluish gray, sea green	**glauc**oma, **glauc**ous
-glen-o	cavity, socket, eyeball	**glen**oid, *Euglena*
-glia-	glue	nero**glia**
-glob-	sphere, all	**glob**ate, **glob**al
-gloch-	point of an arrow	**gloch**idium
-glom-	a ball	**glom**erulus
-gloss-, -glott-	tongue, language	hyo**gloss**al muscle, epi**glott**is
-gluc-o, -glycer-	sweet, sugar	**gluc**ose, **glycer**ol, **glyc**ogen
-glut-	buttock, rump	**glut**eus muscle
-glutin-	glue	**glutin**ous
-glycer-	see -gluc-o	
-gnath-	jaw	**gnath**ostome, **Agnatha**
-gno-	to know	dia**gno**sis
-gnomy	knowledge of judging	physio**gnomy**
-gon-, -gony	seed, reproduction, generation	**gon**ad, oo**gon**ia, glucag**on**
-gon-i	angle	dia**gon**al
-gracil-	slender	**gracil**is muscle, **gracil**e
-grad-	to step	planti**grade**
-gram	a record, something written	cardio**gram,** myo**gram**
-gramin-	grass	**gramin**aceous, **Gramin**eae
-gran-	seed, grain, granular	chloroplast **gran**a, **gran**iform, **gran**ular
-grand-	large	**grand**iose
-graph-	to write	**graph**ic, chromato**graph**y

Root / Prefix / Suffix	Usually Means	As In
-gravi-, -gravid-	heavy, pregnant	**gravid, gravi**tropism
-greg-	flock	**greg**arious
-gul-	throat	**gul**let, **gul**ar
-gust-	taste	**gust**atory
-gutt-	drop, tear	**gutt**ation, **gutt**iform
-gymn-o	naked, uncovered	**gymn**osperm
-gyn-o, -gynec-	a woman, female	**gynec**ologist, **gyn**oecium, poly**gyny**
-gyr-o, -gyrus-	circle, round	**gyr**ate, **gyr**ose

H

Root / Prefix / Suffix	Usually Means	As In
-habenul-	strap	**habenula**
habit-	character, condition	**habit, habit**uation
-habitat-	to live in, to dwell	**habitat, habit**ation
-haem-, hemat-	see -hem-o	
-hal-o	salt, sea	**hal**ophyte, **hal**ogen, **hal**ophobe, **hal**ide
-hal-	to breathe	in**hal**ation
ham-	a hook	**ham**ate bone, **ham**ulose
-hapl-	single, half, simple	**hapl**oid, **hapl**obiontic, **hapl**ostele
-hapt-	to touch	**hapt**en, **hapt**ic
-haust-	to draw out, to drink	**haust**orium
hecto-	one hundred	**hecto**gram
-hedron	a solid shape, having a specified number of sides	poly**hedron**
-heli-o, -hele-o	sun	**hele**oplankton, ***Helianthus,*** **heli**otrophic
-helic-, helix	spiral, coil	**Helic**al, double **helix**
-helminth-	worm	Platy**helminth**es, **helminth**oid
-hem-o, -hemat-, -haem-	blood	**hem**oglobin (**haem**oglobin), **hemat**oma, **hem**ophiliac

Root / Prefix / Suffix	Usually Means	As In
-hemer	day	ep**hemer**al
hemi-	one half	**hemi**sphere, **hemi**branch, **hemi**chordate
-hepat-, -hepar-	liver	**hepat**ic, **hepat**itis, **hepari**
-hept-, -hepta-	seven	**hepta**hydrate
-herb-, -herba-	plant, grass, herb	**herb**ivore, **herba**rium
-hermaphr-o	joined into one body	**hermaphr**odite
-herpat-	reptile	**herpat**ology
-heter-o	other, different	**hetero**sexual, **hetero**gamous, **hetero**genous
-hex-	six	**hex**ose
-hibern-	winter	**hibern**ate
-hidr-	sweat	**hidr**oplankton, **hidr**osis
-hilium-	a trifle	**hilium**
-hipp-	horse	**hipp**uric acid, **hipp**ocampus
-hirud-	leech	**hirud**in, **Hirud**inae
-hist-o	tissue, a web	**hist**ology, **hist**amine, anti**hist**amine
-hod-, -ode	road, way	cat**hode**, an**ode**
-hol-o	whole, entire	**hol**ophyte, **hol**ozoic
-hom-o, -home-o	alike, same, similar, agreeing	**hom**ology, **home**ostasis, **hom**osexual, **hom**ologue
-homo-, -homin-	man	*Homo sapiens*, **homin**oid, **Homin**idae
-homolog-	agreeing, arrangement	**homolog**ous
-horm-	a chain	**horm**ogonium
hormone	to excite, set in motion	**hormone**
-hort-	a garden	**hort**iculture
-host-	friend	**host** organism

Root / Prefix / Suffix	Usually Means	As In
-hum-	earth, ground, soil	**hum**us
-humer-	shoulder	**humer**us, **humer**al
-humor-	fluid, moist, liquid	vitreous **humor, humor**al
-hy-o	Y shaped, U shaped	**hy**oid bone
-hyal-	glass, transparent	**hyal**ine cartilage, **hyal**uronic acid
hybrid	a mongrel, mixed offspring	**hybrid**
-hydr-, -hydr-o	water, fluid (also hydrogen)	**hydr**ation, **hydr**olytic, **hydr**ophytes, carbo**hydr**ate
-hyg-	health	**hyg**iene, **hyg**ienist
-hygr-	moist, wet	**hygr**ometer, **hygr**ophytes
-hymen-	skin, a membrane	**hymen**oid, **hymen**ium
hyp-o	under, beneath, below	**hyp**otonic, **hyp**ocotyl, **hyp**othesis, **hyp**axial
hyp-, hyper-	above, over, excessive	**hyper**tonic, **hyper**tension
-hyph-	a web, something woven	**hyph**a, **hyph**al
-hypn-o	sleep	**hypn**ospore, **hypn**osis
-hyps-o	height	**hyps**odont
-hyster-	uterus, hysteria	**hyster**ectomy, **hysteri**a

I

Root / Prefix / Suffix	Usually Means	As In
-i-	to go	amb**i**ent
-ia	state of, condition of, disease	hysteri**a**
-ia	denoting a taxonomic class	Mammal**ia**
-iac, -ial, -ic, -ian	pertaining to	card**iac,** fil**ial,** gastr**ic,** Cambr**ian**
-iasis, -osis	resulting from, abnormal condition	hypochondr**iasis,** atheroscler**osis**
-iatric, -iatr-	physician, medical treatment	psych**iatric,** ped**iatr**ician
-ible	able to be	flex**ible**
-ic	pertaining to	gastr**ic**

Root / Prefix / Suffix	Usually Means	As In
-ichthy-	a fish	Chondr**ichthy**es
-ician	specialist in	diet**ician,** phys**ician**
-icle	little	aur**icle**
-ics, -tics	art, science of, study of	phys**ics,** gene**tics**
-id, -ide	tending to, pertaining to, member or offspring of	ac**id,** frig**id,** chlor**ide**
-idae, -ida, -id	denoting a taxonomic group (often a family)	Homin**idae,** Homin**id,** Annel**ida**
-ide-	to see	**ide**ology, **ide**ntification
-ide	denoting a chemical compound	polysacchar**ide,** sulf**ide**
-idi-o	distinct, peculiar, to one's self	**idi**opathy, **idi**omorphic
-idium, -idion	little	bas**idium**
-igate	to make, to drive	fum**igate,** irr**igate**
-ign-	fire	**ign**ite
-il	little	fibr**il**
-ile	having the character of	juven**ile**
ile-	twisted	**ile**um
-ili-	the groin, flanks	**ili**ac artery
im-, in-, il-	not	**im**balance, **in**organic, **in**vertebrate, **il**logical
im-	see in-	
-imbib-	to drink	**imbib**ition
-immuni-	except, free, safe	**immuni**ty
-in, -ine	chemical substance	adrenal**in,** humal**in**
in-, im-	in, into	**in**cision, **in**duce, **im**plant
-inae	denoting a taxonomic subfamily	Cerv**inae**
-inal	together	germ**inal**
-incis-	cut into	**incis**ion, **incis**or
-incu-	an anvil	auditory **incus**

Root / Prefix / Suffix	Usually Means	As In
-indigen-	native	**indigen**ous
-indus-	garment	**indus**ium
-induce-	to lead in, to induce	**induc**tion
-ine	having the character of, like	mar**ine,** can**ine,** **pur**ine
-ineum	to excrete, to empty out	per**ineum**
-infer-	underneath, low	**infer**ior vena cava, **infer**ior ovary
inflamm-	to set ablaze	**inflamm**ation
-infloresc-	begin to bloom	**infloresc**ence
infra-	below, beneath	**infra**orbital, **infra**red
-infundibul-	funnel	**infundibul**um
-ing	having the quality of, belong to	molt**ing,** sporl**ing,** divid**ing**
-inguin-	groin	**inguin**al canal
inocul-	engrafted	**inocul**um
-insect-	cut into	**insect**s
insul-	island	**insul**in
-integ-	whole, complete	**integ**rate, **integ**er
-integ-	a covering	**integ**ument
-intestin-	guts	gastro**intestin**al
inter-	between, among, during	**inter**cellular, **inter**breed, **inter**phase
intercal-, -cal-	to insert between	**intercal**ated, **intercal**ary meristem
intra-	within, in, into, inside	**intra**cellular, **intra**venous
intrins-	internally	**intrins**ic
intro-	entrance	**intro**itus
-invert-	to turn upside down, to invert	**inver**sion, **invert**ase
-invol-	enwrapped, rolling inward	**invol**ution
-iole, -ium	little	pet**iole,** arter**iole,** bacter**ium**
-ion	state or process of	adhes**ion,** cohes**ion**

Root / Prefix / Suffix	Usually Means	As In
-ion-o	to go, to wander	**ion**ic
-ious	full of, pertaining to	euphon**ious**, judic**ious**
ipsi-	self	**ipsi**lateral
-ir-, -irid-, -iris-	iris, rainbow	**irid**ocyte, **irid**escent, **iris**
-irrit-, irritab-	anger, to provoke	**irritab**ility
-is-o	equal, same	**is**otonic, **is**ometric, **is**omorphic
-isch-	a deficiency, to hold	**isch**uria
-ischi-	hip	**ischi**um
-ish	belonging to, tending to be	blu**ish**
-isk, iscus	little	men**iscus**
-ism	belief, the process of	vital**ism,** creation**ism**
-ist, -ast	one who does, specialist	biolog**ist,** agon**ist**
-istyli	pillar	amph**istyli**
-ite, -ites	part, belonging to; often used in forming fossil names	som**ite,** trilob**ite,** dendr**ite**
-ition	state of, quality of, process of	imbib**ition**, cond**ition**
-itis	disease, inflammation	dermat**itis,** tendon**itis**
-ity, -ty, -itude	state of, quality of	acid**ity,** long**itude,** aprior**ity**
-ium	part, region, little	cond**ium,** epigastr**ium**
-ium	see -iole	
-ive	tending to, included to	aggress**ive**
-ization	process of	ion**ization,** fertil**ization,** cephal**ization**
-ize	to make, to treat, to turn into	homogen**ize**

J

Root / Prefix / Suffix	Usually Means	As In
-jac-	to lie	ad**jac**ent
-ject-, -jacul-	to throw, dart	in**ject**ion, ej**acul**ate
-jejun-	fasting, empty	**jejun**um
jugular	collarbone	**jugular** vein

Root / Prefix / Suffix	Usually Means	As In
-junct-	to join	dis**junct**ion
-juven-	youth	**juven**ile
-juxta-	near to, by the side of	**juxta**position, **juxta**spinal

K

Root / Prefix / Suffix	Usually Means	As In
-kak-	see -cac-	
-kary-o	the nucleus, nut	pro**kary**otic
-ker-, -kerat-	horn, horny tissue	**kerat**in
kilo-	one thousand	**kilo**calorie, **kilo**metre
-kine-, -cine	movement, moving, move	**kine**siology, **kine**tic
-kym-	see -cym-	

L

Root / Prefix / Suffix	Usually Means	As In
-labi-, -labr-	lip	**labi**al palps, **labi**ate, **labi**um
-labil-, -laps-	to slip, to fall	**labile**
labyrinth-	tortuous passage, maze	auditory **labyrinth, labyrinth**odont
-lacer-	torn, mangled	**lacer**ate
-lacin-	flap, fringe	**lacin**ia, **lacin**ate
-lacr-i, -lachy	tears, weeping	**lacr**imal glands, **lacr**iform
-lact-o	milk	**lact**ose, **lact**ation
-lacun-	small pit, gap, cavity	**lacun**ae
-lagen-	flask	**lagen**a
-lal-	to talk, to speak	rhino**lal**ia
-lamell-, -lamin-	small thin plate	**lamell**a, **lamin**ation
-lan-	wool	**lan**olin
-lapar-	abdomen	**lapar**otomy
-laps-	see -labil-	
-larva-	a ghost	**larva**l
-laryng-	larynx, gullet	**laryng**itis

Root / Prefix / Suffix	Usually Means	As In
-lat-, -later-	broad, wide, side	di**lat**ion, **later**al, **lat**issimus dorsi muscle
-laten-	hidden	**latent** heat
-latic-	juice	**latic**iferous, **latex**, **latic**ifer
-lavat-	to wash	**lavat**ion
-lax-	loose, loosen	**lax**ative
-lecith-	yolk	**lecith**en
-lect-	to choose	se**lect**ive
-lei-	smooth	**lei**odermatous, **lei**otrichous
-lemma	sheath, husk, layer, peel	neuri**lemma**
-lemnisc-	ribbon	lateral **lemnisc**us
-lent-	a lentil, lens-shaped	**lent**iculate, **lent**iform
-lep-	scaly, rough	**lep**idotrich, **lep**idoid, **lep**rosy
-lepsy	seizure, attack	narco**lepsy,** epi**lepsy**
-lept-o	thin, delicate	**lept**oid, **lept**otichous
-ler	eye	ant**ler**
-less	devoid, lacking	stem**less,** hair**less**
-let	small	plate**let**
-leuc-o, -leuk-o	white	**leuc**ocyte, **leuc**oplast, **leuk**emia
-lev-	light weight, to raise	**lev**ator muscle
-lex-	to read, phrase	dys**lex**ia, **lex**icon
-lien-	spleen	**lien**al, **lien**itis
-lig-	to bind	**lig**and, **lig**ase
-ligament-	a bandage	**ligament**
-lign-	wood	**lign**in
-ligul-	a small tongue, small strap	**ligul**ate, **ligule**
-lim-	hunger	bu**lim**ia
-limb-	border	**limb**ic artery
-limin-	threshold	sub**limin**al

Root / Prefix / Suffix	Usually Means	As In
-limn-	a marsh, a lake	**limn**ology
-lin-e	line	**lin**eage, **lin**ear
-lingu-	tongue	sub**lingu**al gland, **lingu**al tonsil
-lip-o	fat	**lip**id, **lip**ase
-liqu-	to be liquid	**lique**faction
-liss-	smooth, agile	**liss**ome
-lith-o, -lite-	rock, stone	**lith**ophilic, **lith**ophyll, **lith**osphere, oto**lith**
-littor-	the seashore	**littor**al zone
-lob-	lobe	**lob**ate, **lob**otomy
-loc-, -loca-, locus	place, position	**locus, loca**tion, gene **loci**, **loc**omotion
-locul-	small chamber	**locule**
-log-, -logue	reason, word, speech	**log**ical, homo**logue**
-logist	a specialist, one who studies	bio**logist**
-logy	study of, science of	bio**logy**
-long-	long	**long**itudinal
-loph-	crest, ridge, tuft	**loph**otrichous
-lor-	thong, strap	**lor**e, **lor**ate
-luc-	light, to shine	trans**luc**ent, **luc**iferase
-lumb-	loin	**lumb**ar region
-lumin-, -lumen	light, an opening	**lumin**escent, **lumen**
-lun-	moon	semi**lun**ar valve, **lun**ate
-lute-	orange-yellow	**lute**in, **lute**inizing hormone
-ly-, -lys-, -lyt-, -lyte, -lyze, -lyst	to loosen, to take apart, to dissolve	ana**lysis**, hydro**lysis**, **lysis**, **lys**osome, cata**lyst,** electro**lyte,** hydro**lyt**ic
-lymph-	water, a clear fluid	**lymph**atic fluid, **lymph**ocyte
-lyt-	see -ly	

Root / Prefix / Suffix	Usually Means	As In
M		
-macer-	to soften	**macer**ation
-macr-o	large, long	**macr**ophage, **macr**androus, **macr**omolecule
-macul-	spot	**macul**ation, **macul**a lutea
-magn-	great	**magn**ification, foramen **magn**um
major	larger	labia **major**
-mal-	bad	**mal**ignant, **mal**aria, **mal**odorous
-malac-	soft	**malac**ology
-mall-	hammer	auditory **mall**eus
-mamm-	a teat, breast	**mamm**ary, **Mamm**alia
man-	thin, scanty	**man**ometer
-man-u	hand, handle	**man**ubrium, **man**ual
-mandibul-	a jaw	**mandibul**ar arch
-manduc-	to chew	**manduc**ation
-mania	madness	klepto**mania,** ego**mania**
-mar-	sea, of the sea	**mar**ine
marg-	edge, border	**marg**inal
-marsup-	purse, pouch	**Marsup**ial
-mas-, -mastic-	to eat, chew	**mas**seter muscle, **mastic**ation
-mass	lump	bio**mass**
-mast-, -maz-	breast	**mast**ectomy, **mast**itis
-mastig-	flagellum, whip	**Mastig**ophora
-masturba-	pollute one's self	**masturba**te
-mater-, -matr-	mother	**mater**nity, **matr**ix
-maxill-	upper jaw	**maxill**ary
maxim-	greatest, large	**maxim**al, **maxim**um temperature

Root / Prefix / Suffix	Usually Means	As In
meat-	passage	auditory **meat**us
-mechan-	contrivance	**mechan**orecepter
medi-, mid-	middle	**medi**al, **medi**um, **mid**-rib, **mid**dle lamella
-medull-	middle, pith, marrow	renal **medull**a, **medull**ary
mega-, megal-	very large, one million	**mega**vitamin, **mega**spore, **mega**phyllous, **megal**ops
-mei-	reduction, less, smaller	**mei**osis, **mei**ospore
-mel-	limb	a**mel**us
-melan-o	black, dark	**melan**ocytes, **melan**in
-melli-	honey	**melli**fluous
-membran-	a coating	**membran**eous
-men-	see -ment-	
-men-, -mens-	a month, monthly, moon	**men**opause, **mens**truation
-mening-	membrane	**mening**itis, **mening**es
-menisc-	a crescent, small moon	**menisc**us, **menisc**oid
-mensal-	table	com**mensal**ism
-ment-, -men-	mind	de**ment**ia, **ment**al
-ment-	condition of, state of being	environ**ment,** nida**ment**al gland, comple**ment**ation
-mere, -meri-, -mer-o	a part of, part	sarco**mere,** myo**mere,** **meri**stem, poly**mer,** iso**mer,** **mer**ocrine
-merist-	divisible	**merist**em
-meryc-	ruminate, chew the cud	**Meryc**oidodon
-mes-o, -mesi-	middle	**meso**derm, **mes**entery, **meso**phyll
meta-	after, middle, between, beyond	**meta**phase, **meta**stasis, **meta**carpal,
-metabol-, meta-	change	**metabol**ic, **metabol**ism, **meta**morphosis
meteor-	high, in air, lofty	**meteor**ic

Root / Prefix / Suffix	Usually Means	As In
-meter, -metry	measure, science of measure	opto**metry,** odo**meter,** thermo**meter,** micro**meter**
-metr	uterus, mother	endo**metr**ium
-metre, -metric	a unit of length	kilo**metre,** micro**metre,** iso**metric**
-mic-	see -mix-	
-micr-o	small, one millionth	**micro**scope, **micro**biology, **micr**androus
-mictur-	urinate, to make water	**mictur**ate
mid-	see medi-	
-migra-	move, wandering	**migra**tion
-milli-	thousandth	**milli**metre
mim-	to imitate	**mim**icry
miner-	ore	**miner**al
minor-, minus-	smaller	labia **minor**
mis-	hate, wrong, incorrect	**mis**ogamy, **mis**opedia
-misc-	mix	**misc**ible
-miss-, -mitt-	to let go, to send	trans**miss**ion, trans**mitt**er
-mito-	thread, web	**mito**chondria, **mito**sis
-mix-, -mic-	intercourse	apo**mix**is
-mne-	to remember	a**mne**sia
-mobil-	to move	**mobil**ity
-mod-	measure	**mod**ulation
-mola-	grind, mill	**molar**
-mole-	a mass	**mole**cule
-mollusc-	soft	**Mollusca (Mollusk**a)
-mon-	to advise	**mon**itor
-mono-	one, single	**mono**saccharide, **mono**mer
-mons-	mountain	**mons** pubis

Root / Prefix / Suffix	Usually Means	As In
-morb-	disease	**morb**ose, ***morbus*** *hungaricus*
-morph-o	form, shape	**morph**ology, a**morph**ous
-mort-	death	**mort**uary, **mort**ality
-morul-	a mulberry	**morul**a
-mot-, -mov-	to move, motion	**mot**ile, oculo**mot**or
-muc-o, -mucus-	mucus, slime, juice	**muc**opolysaccharide, **muc**ilage, **muc**osa
-multi-	many	**multi**fidus muscle, **multi**cellular
-mur-	wall	inter**mur**al, **mur**amic acid
-musc-	moss	**musc**iform, **musc**ology
-muscul-	muscle	intra**muscul**ar
-muta-	change	**muta**tion
-muti-	cut off	**muti**late
-mutu-	reciprocal	**mutu**alism
-my-	too close	**my**opia
-myo-, -myos-	muscle, mouse	**myos**in, **myo**cardium
-mycetes	denoting a taxonomic group (a fungal class)	Oo**mycetes**
-myc-o, -myce-	fungus	**myc**ology, **myce**lium
-myel-	marrow, the spinal cord	**myel**in, **myel**encephalon
-myi-	fly	**myi**ophilae
-myl-	molar, millstone	**myl**ohyoid muscle
-myring-	eardrum	**myring**itis
-myx-o	slime, mucus	**Myx**omycota

N

Root / Prefix / Suffix	Usually Means	As In
-nacr-o	mother-of-pearl	**nacr**eous layer
-nai-	water nymph	**nai**ad
nan-o, nana-	a dwarf	**nan**ometre, **nana**ndrous
-nar-	external nostril	**nar**iform, external **nar**es

Root / Prefix / Suffix	Usually Means	As In
-narc-o	stupor	**narc**otic, **narc**osis
-nas-	nose	**nas**al, **nas**opharynx
-nat-	born	neo**nat**al
-nata-	swimming	**nata**tory
-naupli-	shellfish	**naupli**us
-nav-	ship	**nav**icular bone, **nav**iculoid
-ne-o	new, recent, young	**ne**onatal, **neo**-Darwinism
-nebul-	vaporous	**nebul**ous
-necr-o	dead, a dead body tissue	**necr**otic lesions, **necr**osis
-nect-o	swimming	**necto**pod
-nem-, -nemat-	a thread	**Nemat**oda, proto**nema**
-nephr-o	kidney	**nephr**on, **nephr**idium, **nephr**oid
-nerv-	sinew	in**nerv**ation, **nerv**ous system
-neuro-	nerve	**neuro**n, **neuro**logy
-neutr-	neither one nor the other	**neutr**ality, **neutr**ophilic
-nich-	to nest	**nich**e
-nict-	to wink	**nict**itating membrane
-nid-	nest	**nid**ation, **nid**amental
-nigr-	black	**nigr**oid, **nigr**osine
-nitr-, -nitr-o	soda, salt	**nitr**ogenous, **nitr**ogen, **nitr**ite
-noc-, -nox-	harm	**nox**ious, **noc**ent
-noct-, -nyct-	night	**noct**urnal
-nod-	knot, knob	**nod**ular, **nod**e
-nom-, -nomen-	a name, usage	**nomen**clature
-nom-, -nomy	law, the science of	auto**nom**ic, taxo**nomy**
non-	not	**non**-disjunction
-norm-	order, rule, standard	**norm**ality
-nos-	disease	**nos**ology
-not-o	the back, south	**noto**chord

Root / Prefix / Suffix	Usually Means	As In
-novem-	nine	**Novem**ber
-nox-	see -noc-	
-nuch-	neck	**nuch**al crest
-nucle-	kernel, nut, nucleus	**nucle**us, **nucle**olus
-nud-	naked	**nud**ation, **nud**ibranch
-nulli-	none	**null** hypothesis, **nulli**somy
-nunc-	messenger	inter**nunc**ial nerve
-nurt-	nursing	**nurt**ure
-nutri-	to nourish	**nutri**tion
-nux-	nut	**nuc**iferous, **nuc**ivorous
-nyct-	night	**nyct**anthous
-nymph-	bride, chrysalis, pupa	water **nympha**, **nymph**al

O

Root / Prefix / Suffix	Usually Means	As In
o-, oo-, ov-, ovi-	an egg	**oo**gamous, **oo**gonium, **ovi**duct
ob-	against, toward, opposite	**ob**struction
-obes-	fat	**obes**ity
-oblig-	bound, bind, obliged	**oblig**ate anaerobe
-oblique-	slanting, asymmetrical	**oblique** angle
-obsol-	to wear out	**obsol**ete
-obturat-	to close by stopping up	**obturat**or foramen
-occipit-	the back of the head	**occipit**al bone
-occlu-	shut up	**occlu**sion, **occlu**ded
-oct-o, -oct-a	eight	**oct**opus, **oct**amerous
-ocul-, -ocel-	an eye	**ocul**ar, **ocel**lus
-ode, -odo-	see -oid	
-odont-o, -dont	tooth	hom**odont,** thec**odont,** **odont**oid
-odyn-	pain	my**odyn**ia
-oec-, -oeci-	see -ec-o	

Root / Prefix / Suffix	Usually Means	As In
-oid, -ode, -odo-	like, form, shape, resembling	delt**oid, **nemat**ode, **phyll**ode**
-ol	denoting an alcohol or phenol	ethan**ol, **methan**ol**
-ol-e	oil	**ol**eiferous, **ol**eic acid
-ole, -iole	little	arter**iole, **bronch**iole**
-olecran-	point of elbow	**olecran**on
-olf-, -olfact-	smell	**olfact**ion
olig-o	few, little, scanty	**olig**otropic, **olig**odendrocytes
-om-	shoulder	acr**om**iodeltoid muscle, **om**osternum
-oma	a tumor, swelling	carcin**oma, **glauc**oma**
omasum	paunch, tripe	**omasum**
-ome	mass, abstract group	gen**ome, **bi**ome**
-ombr-o	rain	**ombr**ophyte
-oment-	membrane, caul	greater **oment**um
-omma-	eye	**omma**teum
-omni-, -omnis-	all, every	**omni**vore
-omo-	shoulder	**omo**hyoid
-omphal-	naval, umbilicus	**omphal**itis
-on	a particle	interfer**on, **ex**on, **intr**on**
-onc-o	mass, tumor, swelling	**onc**ology, **onc**ogene
-onchy-	nail	**onchy**cryptosis
-ont-	a being, individual, existing	**ont**ogeny, symbi**ont**
oo-	see o-	
-op-	see -opt-	
-opa-	shady	**opa**que
-oper-	work	**oper**on, **oper**ator
-opercul-	a cover, lid	**opercul**um
-oph-i	snake	**Oph**idia
-opisth-o	behind, the hind part	**opisth**osoma

Root / Prefix / Suffix	Usually Means	As In
-opp-o	opposite, to oppose	**opp**osite
-opsida	denoting a taxonomic group (often a plant class)	Magnoli**opsida**
-opsis	appearance	kary**opsis**
-opsy	view of, examination	aut**opsy**
-opt-, -op-	sight, vision	**opt**ic nerve, **opt**ometry, my**op**ia
-opthalm-	eye	**opthalm**ic nerve, **opthalm**ologist
-optim-	best	**optim**al
-or	state of, result of the act of	tum**or**, err**or**
-or-, -os-	mouth	**or**al, **os**culum
-orb-	circle, eye socket	**orb**ital, **orb**ate
-orch-, -orchi-	testicle	**orchi**d
-ordi-	to begin to weave	prim**ordi**um
-orexis-	appetite	an**orexis**
-organ-	instrument, tool	**organ**ic, **organ**elle, **organ**ogenesis
-ornis-, -ornith-	a bird	**ornith**ology
-orth-o	straight, correct	**orth**opedic, **orth**odontist
-ory	place for, related to	laborat**ory**, olfact**ory**
-ory	see -ose	
-os-, -oss-, -oste-	a bone	**oss**ification, **oste**ocyte
-ose	sugar	dextr**ose**, sucr**ose**
-ose, -ory	full of, resembling	adip**ose**, secret**ory**
-osis	diseased, abnormality, condition of	cystic fibr**osis**, cirrh**osis**, metamorph**osis**
-osm-	smell	an**osm**ia
-osmo-	pushing	**osmo**sis, **osmo**regulation
-oste-o, -oss-	bone	**oste**ocyte, **Oste**ichthyes, **oss**icle

Root / Prefix / Suffix	Usually Means	As In
-osti-o	door, mouth-like opening	**osti**um, **osti**ole
-ostrac-	shell	**ostrac**od
-ot-o	ear	**ot**olith, par**ot**id gland, **ot**ic nerve
-ous, -ious	pertaining to, full of	nerv**ous,** amphib**ious,** nitr**ous**
ovi-, ovum	an egg	**ovum, ovi**duct
-ox-, -oxy-	sharp, acid, oxygen	de**oxy**ribonucleic acid, **oxy**gen, **ox**idation, hyp**ox**emia
-oxys-	quick	**oxy**tocin
-oxida-	to oxidize	**oxida**tion

P

Root / Prefix / Suffix	Usually Means	As In
-pab-	to feed	**pab**ulum
-pachy-	thick	**pachy**dermatous, **pachy**tene
-paed-	child	**paed**omorphosis
-pag-	fixed, firmly united	dorso**pag**us
-palat-	roof of the mouth, palate	**palat**ine tonsils, **palate**
-pale-o, -palae-o	ancient, old	**pale**ontology
-pall-o	to be pale	**pall**or
-palli-	mantle, a Roman cloak	**palli**al, **palli**um
-palm-	hand, flat, broad	**palm**ate
-palp-	to touch, to stroke	im**palp**able, **palp**
-paly-	to scatter	**paly**nology
-pampin-	tendril	**pampini**form
-pan-, -pant-	all, complete	**pan**acea, **Pan**gea, **pan**creas
-panic-	tuft	**pani**culate, **panic**le
-papill-	nipple	**papill**ose
-par-a, -paren-	give birth to, to bear	**paren**t, ovi**par**ous, vivi**par**ous

Root / Prefix / Suffix	Usually Means	As In
para-	beside, near, along side	**para**llel, **para**blast, **para**thyroid
-para-	disorder, diseased condition	**para**lysis, **para**plegic
-pariet-	wall	**pariet**al bone
-part-	part, to divide	bi**part**ite, **part**ition
-parthen-o	a virgin	**parthen**ogenesis, **parthen**ogenic
-partur-	to be in labor	**partur**ition
-pass-	to endure	**pass**ive transport
-patagi-	a border	**patagi**um
-patell-	a dish, plate, small pan	**patell**a bone
-pater-, -patern-, -patr-	father, native land	**patern**ity, **patern**al, allo**patr**ic
-path-o	suffering, disease	**path**ogen, **path**ology
-pathy	treatment of disease	hydro**pathy**
-pause	to cease	meno**pause**
-pect-, -pector-	the beast	**pector**alis muscles
pect-	curdled, congealed	**pect**in
-pectin-	comb, scallop	**pectin**ate, **pectin**eus muscle
-ped-	child	**ped**iatrics
-ped-i, -pes-	foot	bi**ped**al, centi**pede**, **ped**uncle, **ped**icel, ortho**ped**ic, **pes**
-ped-a, -pedag-	instruction, teaching	encylo**pedi**a, **pedag**ogical
-ped-o	soil, ground	**ped**ology, **ped**onic
-pelag-	the sea	**pelag**ic zone
-pell-	skin	**pell**icle
-pellucid-	clear, to shine through	**pellucid**a zone
-pelt-	shield	**pelt**ate
-pelv-	basin	**pelv**ic girdle, renal **pelv**is
-pend-, -pensil-	to hang	**pend**ulate, **pensile**

Root / Prefix / Suffix	Usually Means	As In
-peni-	basket-like	**peni**carp
-penia	want, deficiency	leuko**penia**
-penicill-	a little brush	***Penicillium***
-penn-, -pinn-	winged, feather	**penn**ate, **pinn**ate
-pent-a	five	**pent**ose, **pent**aradial
-pep-, -peps-, -pept-	to cook, digest, soften	**peps**in, **pept**ic acid
per-, perm-	to pass through, by means of	**per**forated, **perm**ease, semi**perm**eable membrane
peri-	around, about, near	**peri**stalsis, **peri**cycle, **peri**cardium, **peri**toneum
perine-	hidden	**perine**al
-period-	a going around	photo**period**ism, **period**icity
-peripher-	outer surface	**peripher**al nervous system
-peristal-	compressing around	**peristal**sis
-periton-	stretched over	**periton**eum
perm-	see per-	
-peron-	pin, fibula	**perone**us muscle
-pes-	see -ped-i	
-petal-	a flower leaf	a**petal**ous, **petal**oid
-petiol-	small leaf	**petiol**ate
-petr-	a rock, stone	**petr**eous, **petr**ify
-pexy	to fasten, to put in place	gastro**pexy**
-pha-, -phe-, -phen-, -phase-	to appear, to show, a stage	**phen**otype, **phen**etic, meta**phase**
-phae-, -phaeos-	dusk, dark	**Phae**ophyta, **phaeos**pore
-phag-o, -phage-	to eat, devour	**phag**ocytosis, eso**phag**us, bacterio**phage**
-phalang-e	a line of soldiers	**phalange**s
-phall-	penis	**phall**ic, **phall**us
phaner-	manifest	**phaner**ogam

Root / Prefix / Suffix	Usually Means	As In
-pharmac-o	drug, poison	**pharmac**ological
-pharyng-	throat, gullet	**pharyng**eal, **pharynx**
-phas-, -phe-	to speak, speech	dys**phasia**, pro**phe**tic
-phase-	a stage, to appear	ana**phase,** inter**phase**
-phe-	see -pha-	
-phell-	cork	**phell**oderm
-phen-o	visible, to show	**phen**etic, **pheno**type
-pher-	to carry, to bear	**pher**omone, peri**pher**al nervous system
-phil-, -philous	to love, have an affinity for	cryo**phil**ic, hydro**phil**ic
-phleb-	vein	**phelb**itis
-phlo-	bark	**phlo**em
-phob-, -phobia	fear, dread, dislike	hydro**phob**ic, claustro**phobia**
-phon-, -phone	sound, voice	**phon**ic
-phore-	to bear, bearer, transmission	chromato**phores**, electro**phore**sis, spermato**phores**
-photo-	light	**photo**synthesis, **photo**taxis
-phra-, -fa-, -fat-	to speak	a**phra**sia, in**fa**nt
-phragm	fence, partition	dia**phragm**
-phren-	mind, diaphragm	**phren**ic nerve, **phren**ology, **phren**itis
-phthi-	to waste away	**phthi**sis
-phy-	to grow	apo**phy**sis
-phyc-	algae, seaweed	**phyc**ology, **phyc**oerythrin
-phyl-	a race, class	**phyl**ogeny, **phyl**um
-phylac-, -phylax-	protection	pro**phylac**tic, ana**phylax**is
-phyll-o	leaf	**phyll**otaxy, chloro**phyll**
-physi-, -physe-	nature, function, growth	**physi**ology
-phyta	denoting a plant division	Conifero**phyta**

Root / Prefix / Suffix	Usually Means	As In
-phyt-o	plant, growth	**phyt**oplankton, gameto**phyte**
-pia-	tender	**pia** mater
-picr-o	bitter	**picr**otoxin, **picr**ic acid
-pigment-	paint	**pigment**ation
-pil-o	hair	**pil**ose, **pil**eate
-pile-	cap	**pile**us
-pin-	pine tree, pinecone	**pin**oid, **pin**eal gland, *Pinus*
-pinn-	feather	**pinn**ate
-pino-	drink	**pino**cytosis
-pir-	pear	**pir**iform lobe, **pir**iformis muscle
-pis-	pea	**pis**iform bone
-pisc-	fish	**pisc**ivorous, **Pisc**es
-pistil-	pestle	**pistil**late
-pithecus	ape	*Ramapithecus*
pituitary	slime, phlegm	**pituitary** gland
-plac-, -plax-	flat plate	**plac**oid scale
-placent-	small flat cake	**placent**a
-plan-	wandering	a**plan**ospore, **plan**kton
-plan-	flat	**plan**ation, **Plan**aria
-plank-	drifting, floating	**plank**ton
-plant-	to plant, a sprout	im**plant**ation
-plas-	development, growth	hyper**plas**ia
-plasm-, -plast-y	molded, formed, skin, membrane	**plasm**a membrane, **plasm**olysis, **plast**id, sym**plast,** cyto**plasm,** blepharo**plasty**
-platy-, -platys-	flat, broad	**platy**dactyl, **platys**ma muscle
-plect-o	twisted	**plect**ostele
-pleg-	a stroke, shock	para**plegic**, **pleg**etropism

Root / Prefix / Suffix	Usually Means	As In
-plei-o, pleist-o	more, many	**plei**otropism, **plei**omorphy
-pleur-	the side, rib	**pleur**al cavity
-plex-	to braid, interweave, network	apo**plex**y, **plex**us, **plex**iform
-plic-	to fold, folded	**plic**ation
-ploid, -ploidy	set of chromosomes, multiple of	di**ploid,** ha**ploid,** poly**ploidy**
-plum-	feather	**plum**ule
-plur-	more, many	**plur**ilocular
-pnea-, -pneum-o	breath, air, lung	tachy**pnea** **pneum**onia, **pneum**atocyst
-pod-	a foot	**pod**ite, para**pod**ia, arthro**pods**
-poie	create, to make, production	hemato**poie**sis
-poikil-o	varied, irregular	**poikilo**thermal
-pol-	pole, end of an axis	de**pol**arization
-polio-	gray matter	**polio,** **polio**myelitis
pollen	fine flour, dust	**pollin**ation
pollut-	defiled	**pollut**ion
poly-	many, much	**poly**mer, **poly**peptide
-polyp-o	small growth	**polyp**
pom-	fruit	**pom**iferous
-pon-	to work, to toil	hydro**pon**ics
pons	bridge	**pons** Varolli
-popul-	people	**popul**ation
-por-	opening, passage, channel	**por**ous, blasto**pore**, **Por**ifera
-porphyr-	purple	**porphyr**in
-port-	gate	hepato**port**al
-pose	to rest	decom**pose**
-posit-	to place, to put	ap**posit**ional growth

Root / Prefix / Suffix	Usually Means	As In
post-, poster-	after, behind	**poster**ior
-pot-	to be powerful	toti**pot**ent, **pot**ential energy
-pound-	to put, place	com**pound**
-potam-	river, stream	**potam**ic plankton, **potam**ophyte
pre-	before, in front of	**pre**maxillary bone, **pre**synaptic junction, **pre**adaptation
-pred-	to prey upon	**pred**aceous
-prehend-	to seize	**prehens**ile
-presby-	old	**presby**atrics, **presby**opia
-pri-, prion	jagged, like a saw	**pri**odont
prim-, primord-	first	**prim**ate, **prim**itive, **primord**ial
pro-	before, in front, earlier than, prior to	**pro**phase, **pro**lactin
-prob-	to test, good	**probe**
-proct-o	the anus, rectum	**proct**ologist, **proct**al fin
promo-	to move forward	**promo**ter
-pron-	inclined, face down, bend forward	**pron**ator
-propri-	one's own	**propri**oceptor
pros-	toward, in addition to	**pros**thetic
prostat-	one who stands before	**prostate** gland
prostrat-	lying face down	**prostrate**
-prot-o, -prote-	first, original, primitive	**proto**plast, **proto**zoan, **proto**xylem, **prote**in, **Prot**ista
-proxim-	nearest	**proxim**al
-prur-	to itch	**prur**itus
-pseudo-	false	**pseudo**science, **pseudo**stratified
-psil-o	naked	**Psilo**phyta

Root / Prefix / Suffix	Usually Means	As In
psor-	to itch, scurvy	**psor**iasis
-psych-o	mind, soul	**psych**ology, **psych**
-psychr-o	cold	**psychr**ophilic
-pter-o	fern	**Ptero**phyta
-pteryg-o	wing, fin	**pteryg**oid process
-ptil-	feather	**ptil**inum
-pty-	to spit	**pty**alin, **pty**lism
-pub-, -pubes-	a young adult, to grow hair	**pub**erty, **pub**is, **pubes**
-pulmon-	a lung	**pulmon**ary
puls-	beat, drive, push	**puls**ation, **puls**ate
-pulver-	dust	**pulver**aceous
-pulvin-	cushion	**pulvin**us
-punct-	a prick, a sting	**punct**ure, acu**punct**ure
-pung-	to prick, point	**pung**ent
-pupa-	doll, girl	**pupa**, **pup**ation, **pup**il
-pur-	pure, clean	**pur**ification, **pur**ine
-put-	to think, to prune	com**put**er, am**put**ate
putamen	shell, pod	**putamen**
-putr-	rotten	**putr**efaction
-py-	pus	**py**uria, **py**in, **py**ogenesis
pycn-o	thick	**pycn**iform
-pyel-	a trough	**pyel**itis
-pyg-	rump, buttocks	**pyg**ostyle
-pyl-, -pyle	gate, entrance	micro**pyle**, **pyl**orus
-pylor-	a gatekeeper	**pylor**ic valve
-pyr-, -pyret-	fire, fever	anti**pyret**ic, **pyr**exia
-pyren-	the stone of a fruit	**pyren**oid

Q

Root / Prefix / Suffix	Usually Means	As In
-quadr-	four, to make square	**quadr**uped, **quadr**ate bone

Root / Prefix / Suffix	Usually Means	As In
-quant-	how much	**quant**itative
-quart-	fourth	**quart**er
-quartern-	by fours	**quartern**ary
-quasi-	nearly, almost	**quasi**-science
-quiesc-	quiet, dormant	**quiesc**ent center
-quint-	fifth	**quint**ary

R

Root / Prefix / Suffix	Usually Means	As In
-rabd-, -rhabd-	rod	**rabd**oid, **rhabd**ite
-racem-	cluster of berries, bunch	**racem**ose
-rach-i, -rhach-i	spine, backbone	**rach**is, **rach**illa
-radi-	a spoke, ray, radiating	**radi**al symmetry, **radi**oactive
-radic-	a root	**radic**al, radish
-radul-, -ras-	to scrape	**radul**a, **ras**p
-ram-	branch	**ram**ate, **ram**ify
-raph-, -rhaph-	longitudinal fissure, seam	**raph**e
re-	again, back, to act again	**re**adsorption, **re**tractor muscle, **re**combinant DNA
-rect-	straight	**rect**um, ar**rect**or muscle
-recept-	to receive	chemo**recept**or
recess-	receding	**recess**ive allele
reflex-	bend back	**reflex** arc
-reduct-	reduced	**reduct**ionism
-regres-	a retreat	**regres**sion
-regul-	rule	osmo**regul**ation
-ren-	kidney	**ren**al, **ren**iform, **ren**in
-replic-	to answer to	**replic**ation
-repres-	to keep back	**repres**sor molecule
-rept-, -reptil-	creep, crawl	**reptil**ian

Root / Prefix / Suffix	Usually Means	As In
-resol-	untying, to unbind	**resol**ution
-respir-	breathing, to breathe	**respir**ation
-rest-	rope	**rest**iform body
-reti-	net, network, latticed	**reti**culate, **reti**nal, endoplasmic **reti**culum
-retract-	to pull back	**retract**or muscle
retro-	backward, behind	**retro**gression, **retro**virus
-revol-	to roll back	**revol**ute
-rhabd-	see -rabd-	
-rhach-i	see -rach-i	
-rhaph-	see -raph-	
-rhe-, -rheum-	to flow, current	**rheum**atic fever, **rhe**otropism
-rhin-	nose	**rhin**itis, **rhin**oceros
-rhiz-o, -rhyz-o	a root	**rhiz**ome, **rhiz**oid
-rhod-	red, a rose	**Rhod**ophyta
-rhomb-	parallelogram with oblique angles and unequal sides	**rhomb**oideus muscle, **rhomb**oid
-rib-o	currant (also in reference to a 5-carbon sugar common in some fruit sugars)	**rib**oflavin, **rib**ose
-rig-	stiff, stiffening	**rig**ormortis
-rim-	crack, a cleft	**rim**iform
-rod-	to gnaw	**rod**ent, e**rode**
-rostr-	beak, snout	**rostr**um
-rot-	wheel, round	**rot**uliform, **rot**ate
-rrhaphy	sew, suture	cardio**rrhaphy**
-rrhea, -rrhag-	flow, discharge	gono**rrhea,** dia**rrhea**
-(r)rhythm-	measured motion	a**rrhythm**ia
-rub-, -rubr-	red	bili**rub**in, **rub**ella
-rudiment-	first beginning, unformed	**rudiment**ary

Root / Prefix / Suffix	Usually Means	As In
-rug-	a wrinkle, fold	gastric **rug**ae, **rug**ate
-rumen-	throat	**rumen**
-rumin-	chew, to chew the cud	**rumin**ate
-rupt-	to break, to burst	**rupt**ure, dis**rupt**

S

Root / Prefix / Suffix	Usually Means	As In
-sacc-	sac, bag	auditory **sacc**ule, **sacc**ate
-sacchar-	sugar, sweet	poly**sacchar**ide, **sacchar**in
-sacr-	sacred	**sacr**al nerve, **sacr**um, **sacr**olumbar
-sagitt-	arrow	**sagitt**al plane
-sal-	salt	**sal**ine
-sali-	leaping	**sali**ent
-saliv-	spittle	**saliv**ary gland
-salpin-	trumpet	**salpin**x, meso**salpinx**
-salt-a	to leap, to jump	**salt**atory, **salt**ation, **salt**igrade
-san-	healthy	**san**itary
-sangui-	blood	**sangui**nivorous
-sapr-o	rotten, putrid	**sapr**ophytic, **sapr**ozoic
sarcin-	package	***Sarcina***
-sarc-o	flesh, muscle	**sarc**omere, **sarc**oma
-sartor-	tailor	**sartor**ius muscle
-satur-	to fill up	**satur**ated fatty acids
-saur, -saurus	lizard	dino**saur,** *Tyranno**saurus***
-sax-	rock	**sax**igenous
-scab-	rough, scurvy	**scab**rous
-scal-	ladder	**scal**a vestibule, **scal**ariform
-scalen-	a triangle with three unequal sides	**scalen**us muscle
-scand-, -scans-	to climb	**scand**ent

Root / Prefix / Suffix	Usually Means	As In
-scaph-	bowl, boat	**scaph**oid
-scat-	dung	**scat**ology
-schist-, -schiz-o	split, cleave, deeply divided	**schist**osome, **schiz**ome, **schiz**ocarp
-sci-, -ski-, -skia-	shade	**sci**ophyte, **skia**gram
-sci-	knowledge	**sci**ence
-sciss-	to cut	ab**sciss**ion
-scler-	hard	**scler**eid crystals, **scler**enchyma, athlero**scler**osis
-scoli-o	bent, crooked	**scoli**osis
-scop-, -scopy	to view, examine visually	micro**scope**, micro**scopy**, endo**scop**ic surgery
-scop-	broom, brush	**scop**ulate
-scrot-	pouch, hide	**scrot**um, **scrot**iform
-scut-	a dish, a shield	**scut**ellum
-scyph-, -skyph-	cup	**Scyph**ozoa, **scyph**oid
se-, sed-	going, away, apart	**se**crete, **sed**ition
-seb-	grease, tallow	**seb**aceous gland
-sec-, -seg-, -sect-	to cut	**sec**tion, **seg**ment, **sect**or
-secret-	to separate, to sever	neuro**secret**ion
-sed-, -sess-	to sit, to settle, without a stem	**sed**imentary, **sess**ile
-segment-	a piece cut off	**segment**ation
-segreg-	to separate	**segreg**ation
-sell-	saddle, seat	**sell**aeform
-semen-, -semin-	seed, sowing	**semen**, **semin**al, in**semin**ation
semi-	one half	**semi**permeable membrane, **semi**lunar valve
-sen-, -senesc-	old	**sen**ile, **senesc**ence
-sens-, sent-	feeling, to perceive	**sens**ory

Root / Prefix / Suffix	Usually Means	As In
-sep-	covering	**sep**al
-sep-, -sept-	barrier, wall, partition	myo**sept**um, a**sept**ate
-sept-, -septem-	seven	**sept**imetry, **Septem**ber
-sept-, -sep-	rotten, infected	**sep**sis, **sept**ic, **sept**icemia, anti**sept**ic
sere-	to put in a row	**sere**
ser-o	the watery part of fluids	blood **ser**um, **ser**ology
-serr-, -serra-	a saw, sawtooth	**serra**ted
-sesam-	seed of the sesame plant, granular	**sesam**oid bone
-sess-	see -sed-	
-set-, -seta-	a bristle	**setae**
-sex-	six	**sex**tuplets, **sex**tant
-sial-	saliva	glyco**sial**ia
-sicc-	dry	de**sicc**ation
-sigm-	S-shaped	**sigm**oid curve, **sigm**oid colon
-silic-	a flint	**silic**ious
-silv-, -sylva-	woods, trees	**sylva**culture
-simil-, -simul-	like	as**simil**ate, **simul**ation
-sinus-	curve, hollow, cavity	**sinus**oid, **sinus** venous
-sinistr-	left	**sinistr**al
-siph-o	pipe	**siph**on, **siph**onous
-sis, -sia, -sy	process, action, the act of, condition of	mito**sis,** amne**sia,** ecsta**sy**
-sit-	grain for food, food	**sit**otoxin, **sit**ology
-skelet-	a dried hard body, mummy	**skelet**al system
-skia-	see -sci-	
-soci-, -socia-	a companion, fellow, to join together	**soci**al, **soci**ety, dis**socia**te
-sol-	sun	**sol**ar

Root / Prefix / Suffix	Usually Means	As In
-sole-	sandal, slipper	**sole** of the foot, **sole**us muscle, **sole**aeform
-soli-	alone	**soli**tary
-solv-, -solut-	dissolve, loosen	**solv**ent, **solut**ion
-som-e, -somat-	body	chromo**some**, **somat**ic, ribo**some**
-somn-	sleep	in**somn**ia, **somn**iferous
-son-	sound	ultra**son**ic
-soph-	wise	philo**soph**y
-sorb-	suck in	ab**sorb**, ad**sorb**
sorus	a heap	fern **sorus**
-spasm-	to stretch, convulsion, tension	muscle **spasm**
-speci-	a kind or sort, special	**speci**es, **speci**men
-spectr-, -spec-	to look, appearance	**spectr**um
-sperm-, -spermat-	seed	**spermat**ocyte, angio**sperm**
-sphen-o	wedge	**sphen**oid, **Sphen**ophyta
-spher-o	sphere	**spher**oid
-sphing-	squeeze, to bind tightly	**sphin**cter muscle
-spic-	spike, head of grain	**spic**ate, **spic**ule
-spin-	thorn, spine	**spin**ose, **spin**alis muscle
-spir-	coil, twisted	**spir**al, **spir**ochete
-spira-	to breathe	tran**spira**tion, re**spira**tion
-spirac-	air hole	**spirac**le
-splanchn-o	viscera, entrails, viscus	**splanchn**ocranium
-splen-	spleen	**splen**ic artery
-spor-	seed, spore	**spor**angium, **spor**ophyte
-spum-o	foamy	**spum**ose, **spum**escent
-spur-	false	**spur**ious
-squam-o	scale	**squam**ous epithelium

Root / Prefix / Suffix	Usually Means	As In
-sta-, -stat-	to control, to stop, standing still, stationary	**stat**ic, **stat**olith, epi**stas**is, homeo**stasis**
-stal-, -sten-o	a contraction, compression, a narrowing	peri**stal**sis, **sten**osis, angio**sten**osis, ***Stentor***
stamen-	anything standing upright	**stamen**
-stap-	stirrup	**stap**es bone
-staphyl-o	a cluster of grapes	***Staphylo**coccus*
-stear-, -steat-	fat, tallow	**stear**ic acid
-stel-	a pillar	**stele**
-stell-	starry	**stell**ate
-sten-o	see -stal-	
-steph-	crown-like	**steph**anous
-sterc-	dung	**sterc**oricolous
-ster-	hard, firm, solid	**ster**oids
-stereo-	three-dimensional, solid	**stereo**scopic vision
-stern-	the breast, chest	**stern**um
-steth-o	chest	**stetho**scope
-sthen-	strength	cali**sthen**ics
-stict-o	spotted	**stict**opetalus
-stigm-	mark, point, spot	**stigm**a, **stigm**atism
-stip-	stalk	**stip**ule
-stol-, -stal-	to send, to contact, to shoot	systole, peri**stal**sis
-stom-, -stome-	mouth, opening	**stom**ata, deutero**stome**
-stomy	the making of an opening	arthro**stomy**
-strab-	to squint	**strab**ismus
-strat-	a layer	**strat**um, **strat**ification, sub**strate**
-strept-o, -streph-	bent, twisted chains	***Strepto**coccus*
-stria-	furrow, groove, streaked	**stria**tion
-strict-	to draw tight	con**strict**ion
-strobil-	cone	**strobil**us

Root / Prefix / Suffix	Usually Means	As In
-strom-	anything spread out, bed, mattress, bedding	**stroma**
-struct-	to build	ultra**structure**
-styl-	pillar, stalk	**style**, **styl**oid, **styl**omastoid
sub-, sus-	beneath, below, under, up from under	**sub**mandibular, **sus**pend, **sub**littoral, **sub**alpine
substrat-	strewn under	**substrate**-level phosphorylation, **substrate**
-succ-	juice, sap	**succ**ulant, **succ**iferous
-success-	to follow	ecological **success**ion
-sucr-	sugar	**sucr**ose
-sud-	to sweat	**sud**ation
-sulc-	a furrow, groove	**sulc**ate
-sum-	to use	con**sum**er
super-, supra-	above, over	**super**ior vena cava, **supra**spinatus muscle, **supra**littoral
superfici-	surface	**superfici**al
-supin-	to bend backward	**supin**ator muscle
-sur-	above, beyond	**sur**face
sus-	see sub-	
-sutur-	to sew	**sutur**e
-sy	see -sis	
-sylva-	see -silv-	
sym-, syn-	together with, a union, junction, to live with	**syn**thesis, **sym**patric, **sym**biosis
-symmet-	measured together	a**symmet**rical
-sympath-	of like feelings	neuro**sympath**etic
-synapt-	clasp	**synapt**ic
-synerg-	work together	**synerg**estic
-syring-	tube	**syring**e

Root / Prefix / Suffix	Usually Means	As In
-system-	a composite whole	**system**ic artery, organ **system**
-systol-	a contraction	**systole**

T

Root / Prefix / Suffix	Usually Means	As In
-tabl-	a board, plank, tablet	**tabl**oid, **tabl**e bones
-tach-, -tachy-	swift, fast	**tachy**cardia
-tact-, -tang-	touch	**tact**ile, **tang**ential
-taeni-	ribbon	**taeni**oid, **taeni**a
-tagm-a	arrangement, order	**tagm**osis
-tal-	ankle	**tal**us bone, **tal**on
-tang-	to touch	**tang**ential section
-tape-	carpet	**tape**tum
-tars-	in step	**tars**optosis, meta**tars**al
-tax-o	arrangement, order, rank	**taxo**nomy, phyllo**taxis**
-taxis	ordered movement	chemo**taxis,** geo**taxis**
-tech-	techniques	bio**tech**nology
-teg-, -tectum-	a covering, roof	in**teg**ument, **tectum**
tel-, tele-	end, completion, distant, far	**tel**ophase, **tele**ost, **tele**metry
-tela-	web-like membrane	**tela** choroidea
tele-	see tel-	
-tem	setting in order	sys**tem**
-temp-	period of time	**temp**oral isolation
-tempor-	the temples	**tempor**al bone
-ten-, -tend-, -tends-	stretch, extend	hyper**tens**ion, **tend**on, **tend**ril, neo**ten**y
-ten-, -tent-	to hold, feeler	**tent**acle, **ten**ure, **ten**acious
-tentori-	a cover, tent	**tentori**um
-tenu-	slender	**tenu**ous, **tenu**ate
-ter	means of, place of	ure**ter,** sphinc**ter**

Root / Prefix / Suffix	Usually Means	As In
-terat-o	monster	**terat**ology
teres	smooth, rounded	**teres** major muscle
-termin-	an end, limit	de**termin**ate growth
-terr-, -terra-	the earth, land	**terr**estrial, **terra**rium
-territor-	domain	**territor**ial
-terti-, -ter-	three times, third	**terti**ary
test	a tile, shell	**test**
testis	testicle	**testis**
-tetan-	stretched	**tetan**us
-tetr-, -tetra-	four	**tetr**ad, **tetra**hedral
-text-	weave	**text**ure, **tissue**
-thalam-	bed, inner chamber	**thalam**us
-thall-	a young shoot, twig	**thall**ophyte, **thall**us
-thanat-	death	**thanat**oid
-the-, -thes-	to put, to place, setting in order	syn**thes**is, hypo**thes**is
-thec-	cavity, receptacle, case	hypo**thec**a, epi**thec**a
-thel-	nipple, female	epi**thel**ium
-theor-	to look at	**theor**y, **theor**em
-ther-	wild beast	proto**ther**ia
-ther-o	summer	**ther**ophyte
-therap-	nurse, care	chemo**therapy**, **therap**eutic
-therm-o	heat, warm	**therm**ophile, homeo**therm**, hypo**therm**ia
-thes-	see -the-	
-thi-	sulfur	**thi**obacteria
-thigm-	a touch	**thigm**otropism
-thorac-, thorax	a breastplate, chest	**thorac**ic cavity, **thorax**
-thromb-o	a clot or lump	**thromb**osis
-thylak-	pouch, bag	**thylak**oids
-thym-	mind	**thym**us

Root / Prefix / Suffix	Usually Means	As In
-thyr-	an oblong shield, door	**thyr**oid gland
-tic, -stic	pertaining to the process of	synthe**tic**, diagno**stic**
-tics	study of, science of	gene**tics**
-tion	process of, action of	evolu**tion**
-toc-	birth	oxy**toc**in
-toler-	to bear	**toler**ance
-tom-, -tomy	to cut, section, cutting	micro**tom**e, dia**tom**, a**tom**, appendec**tomy**
-ton-, -tonis-	condition of	iso**ton**ic, **ton**icity
-ton-	something stretched	peri**ton**eum
-top-o	a place, position	**topo**graphy, iso**tope**
-tor-	a bulge, swelling	**tor**us
-tort-, -tors-	to twist	**tort**uous vein, **tors**ion
-tot-, -toti-	all, so many	**toti**potent
-tox-i	a poison	**tox**in, **tox**icology, anti**tox**in
-trabecu-	small beam, small plate	**trabecu**lar bone
-trache-	windpipe, rough	**trache**a, **Trache**ophyte
-tract-	to drag, to draw	**tract**ion, pro**tract**or
trans-	across, through, beyond	**trans**verse, **trans**cription
-transpos-	to reverse the order of	**transpos**ons
-trapez-	a figure of four unequal sides	**trapez**oid
-trauma-	wound	**trauma**tize
-trem-	hole	Mono**trem**ata
tret-	to perforate, to pierce	**tret**aceous
tri-	three	**tri**ploid, **tri**sect, **tri**ceps
-trib-, -trip-	to rub	**trib**ium
-trich-o	hair	**trich**ome
-tripsy	crushing	nephrolitho**tripsy**
-troch-o	wheel, disk	**troch**ophore
-trochle-	a pulley	**trochle**a

Root / Prefix / Suffix	Usually Means	As In
-trop-, -tropy	to turn, to change	**trop**ic, gonado**tropin**, geo**tropism**
-troph-o, -trophy	nutrition, to feed	**troph**ic level, auto**trophy**, **troph**oblast
-trum, -tr-	result of the act of, means of	spec**trum**
try-	wear down, rub out	**try**psin
-tuber-	a knot, knob, a swelling	plant **tuber**, **tuber**culosis
tubul-	a small tube	**tubul**ar
-tum-	to swell	**tum**or, **tum**id
-tunic-	a cloak or covering	**tunic**a
-turb-	whirl, a spinning object	**turb**ine, **turb**inate bone
-turg-	swell, inflated	**turg**id
-tuss-	cough	per**tuss**is
-ty	state or condition of being	immuni**ty**, sani**ty**
-tycho-	free-floating	**tycho**plankton
-tyl-	callus	**tyl**osis
-tympan-	a drum, tambourine	**tympan**ic membrane, **tympan**ic bulla
-typ-	form	geno**type**, pheno**type**
-typhl-	blind	**typhl**osole

U

Root / Prefix / Suffix	Usually Means	As In
u-, un-	not	**un**scientific, **un**saturated
-ubiqu-	everywhere	**ubiqu**itous
-ule, -ula	little	spic**ule**, ven**ule**, molec**ule**
-ulent, -ulous	full of, disposed to, tending to	vir**ulent**, pur**ulent**
-ulig-	marshy	**ulig**nous
ultra-	beyond the normal	**ultra**structure, **ultra**violet
-um	structure, tissue, thing	cerebr**um**

Root / Prefix / Suffix	Usually Means	As In
-umbel-	see -umbr-	
-umbil-	the navel	**umbil**ical cord
-umbo-	a projecting knob, convex surface	molluscan **umbo, umbo**nate
-umbr-, -umbel-	shade, shadow, sunshade	**umbr**ophilic, **umbel**
un-	see u-	
un-, uni-	one	**uni**cellular, **uni**nucleate
-unc-	hook	**unc**iform
undulat-	risen like waves	**undulat**ions
-ungul-	hoof, nail	**ungul**ate
-uper- (-hyper-)	over, above, excessive	**hyper**tonic
-upo- (-hypo-)	under, beneath, below	**hypo**tonic
-ur-o, -uri-	tail	**ur**ocele, an**uria**
-ur-o	urine	**ur**ea, **ur**acil, **ur**ophile
-ure	act of, result of the act of	rupt**ure,** fract**ure**
-ured-	blight	**ured**iopspore
-ureth-	canal	**ureth**ra
-uria	condition of the urine	hemat**uria,** melan**uria**
-us	person, individual, thing	uter**us,** nucle**us,** epididym**us**
-uter-	womb	**uter**us
-utr-	leather bag, skin bottle	**utr**icule, **utr**iform
-uve-, a	grape	**uve**al

V

Root / Prefix / Suffix	Usually Means	As In
-vacu-	empty	**vacu**um, **vacu**ole
-vaccin-	of a cow	**vaccin**e, **vaccin**ation
-vacil-	swaying	**vacil**lation
-vag-	wandering, undecided	**vag**us nerve, **vag**iform
-vagin-	sheath	**vagin**a, in**vagin**ation
-val-	to be strong, to be well	**val**ence electrons

Root / Prefix / Suffix	Usually Means	As In
-vall-	to surround with a rampart	**vall**ate papilla
-valv-	a folding door	**valve**
-vari-	smallpox, pustule, change	**vari**ole, **vari**ation
-varic-	a dilated vein, ridges	**varic**ose vein
-vas-o	vessel, duct	**vas** deferens, **vaso**dilation
vascul-	a small vessel	**vascul**ar bundle, cardio**vascul**ar
-vast-	large area, immense	**vast**us lateralis muscle
vect-	carry	**vect**or
-veget-	to enliven	**veget**ative
-vel-	veil, covering	**vel**um
-velop-	to wrap	nuclear en**velop**e
-ven-	a vein	**ven**ous, **ven**ation
venere-	pertaining to Venus, coitus	mons **vener**is, **venere**al disease
ventilat-	to fan	**ventilat**ion
-ventr-	the under side, belly	**ventr**al, **ventr**icle
verg-	to incline	con**verg**ent evolution
-verm-	worm	**verm**iform, **verm**is
-vern-	sloughing	**vern**ation
-vernal-	spring	**vernal**ization
-vertebra-	a joint	**vertebra**te, in**vertebra**te
-vers-	to turn	in**vers**ion
-vesicul-	small bladder	**vesic**le, **vesicul**ar gland
-vestibul-	entrance chamber, porch	**vestibul**e
-vestig-	trace, a footprint	**vestig**ial organ
-via-	life	**via**ble
-vibr-	to agitate	**vibr**issa, **vibr**ate
vig-	strong	**vig**or
-vill-	hairy, tuft of hair	**vill**ous, **vill**i, **vill**iform

Root / Prefix / Suffix	Usually Means	As In
-vir-, -virul-	poisonous slime	**virus**, **virulent**
-virg-	a virgin, maiden	**virg**in
-viril-	man	**virile**
-vis-, -vid-, -visu-	to see	**vis**ion, vi**vid**, **visu**al
-visc-, -viscer-	internal organs, entrails	**viscer**al, e**viscer**ated
-visc-	clammy, sticky	**visc**ous
-vit-a	life	**vit**alism, **vita**min
-vitell-	the yolk of an egg	**vitell**ine gland
-viti-	pertaining to a vine	**viti**culture
-vitre-	glass	*in vitro*, **vitre**ous body
-viv-	alive, living	*in vivo*, **viv**iparous
-volunt-	of one's free will	**volunt**ary muscle
-volv-, -volut-	to roll, to turn	in**volute**, **volut**in, e**volut**ion
vomer	plowshare	**vomer** bone
-vor-	to eat	omni**vor**ous, **vor**acious, insecti**vor**ous
-vort-, -vert-	to turn	**vort**ex, di**vert**iculum
-vul-	pluck, pull, tear	re**vul**sion
vulv-	a covering	**vulv**a

X

Root / Prefix / Suffix	Usually Means	As In
-xanth-o	yellow, yellow-brown	**xanth**ophyll
-xen-	stranger, host	**xen**ophobia
-xer-	dry	**xer**ophytic, **xer**ic
-xiph-	sword	**xiph**ihumeralis muscle, **xiph**oid
-xyl-o	wood	**xyl**em, **xyl**ose

Y

Root / Prefix / Suffix	Usually Means	As In
-y	state of, condition of, process of	energ**y**, agon**y**, atroph**y**
-yl	a chemical radical group	eth**yl**, but**yl**, carbox**yl**

Root / Prefix / Suffix	Usually Means	As In
-ysis	growth	ecd**ysis**, diaph**ysis**
Z		
-zation	state of	cephali**zation**
-zea-	a kind of grain	***Zea** mays*, **zea**tin
-zo-, -zoo-	an animal, living, being	**zoo**logy, proto**zo**an
-zoid	animal-like	spermato**zoid**
-zon-, -zona-	a belt, girdle	**zon**ation
-zyg-o	yoke, paired together, union	**zyg**ote, homo**zyg**ous
-zym-	ferment, enzyme, leaven, yeast	en**zym**e, **zym**ogen granules

3

Glossary of Descriptive Terms

Colors and Related Qualifying Terms

Color / Qualifier	Root Word / Prefix	As In
ashen	ciner-	in**ciner**ation
	tephr-o	**tephr**osis (to burn to ashes)
banded	fascia-	**fascia, fascia**ted
black	atr-i	**atr**ium
	coraci-	**coraci**form
	melan-o	**melan**ocyte
	nigr-i	**nigr**oid
	carbon-	**carbon** (coal black)
	eben-	**ebe**neous (ebony black)
	ur-	**ur**eaceus (charred-black)
blue	cerule-	**cerule**an (sky blue)
	cyan-	**cyan**ic
bronze	aene-	**aene**ous
	chalc-	**chal**ceous
brown	bad-i	**bad**ious (chestnut)
	brunn-	**brunn**ette (dark brown)
	castan-	*Castanea* (chestnut genus)
	fulv-	**fulv**ate (brown-yellow)
	fusc-	in**fusc**ate (dusky-brown)
	phae-o	**Phae**ophyta
	tabac-	**tabac**inus (tobacco-brown)

Color / Qualifier	Root Word / Prefix	As In
color	chrom-o	**chromo**plast, **chromo**some
	chromat-o	**chromat**ophore
copper	aer-	**aer**uginous, **aer**eus
dark	amaur-	**amaur**osis
	calig-	**calig**inous (obscure, dark)
	maur-o	**maur**eus
	opa-	**opa**que (shady)
earth	terr-	**terr**eus, **terr**estrial
flesh	carn-	**carn**eus
glassy, transparent	hyal-	**hyal**ine cartilage
	pellucid-	**pellucid**
	vitr-	**vitr**eous
glossy, shiny	nitid-	**nitid**eous
gold	aur-	*Aurelia*, **aur**ous
	chrys-o	**Chrys**ophyta
gray	canes-	**canes**cent
	glauc-o	**glauc**oma (bluish-gray)
	gris-	**gris**on
	incan-	**incan**ous (hoary)
	polio	**polio**encephalitis
	fum-	**fum**ous
	tephr-	**tephr**eus (ashy-gray)
green	chlor-o	**chloro**phyll
	virid-i	**virid**ans, **virid**ulus
	glauc-o	**glauc**escent (gray-green)
	smaragd-	**smaragd**ine (emerald-green)
iron	ferr-	**ferr**ous
lead	plumb-	**plumb**eous
	molybd-	**molybd**enum
	liv-	**liv**id
milky	lact-	**lact**escence
mottled	varieg-	**varieg**ated (partly-colored)
orange	auranti-	**auranti**ous
	cirrh-o	**cirrh**osis (tawny)
	croc-o	*Crocus* (saffron)
pale	pall-	**pall**enate, **pall**or

Color / Qualifier	Root Word / Prefix	As In
purple	porphyr-	**porphyr**in
	purpur-	**purpur**ic
rainbow	irid-	**irid**escent
	iris-	**iris**
red	cardinal-	**cardinal**
	erythr-o	**erythr**ocyte
	ign-	**ign**eus (fiery-red)
	puni-	**puni**ceus (reddish-purple)
	pyr-o	**pyr**etology
	rub-i	**rub**ella (German measles)
		rubicundus (bluish-red)
	rubr-	**rubr**ic
	ruf-i	**ruf**escent
	rut-	**rut**ilant (glowing-red)
	sangui-	**sangui**ferous
rose	rhod-o	**rhod**odendron, **rhod**ochrous
	ros-	**ros**eate
	eos-	**eos**in
rust	ferrugin-	**ferrugin**ous
scarlet	coccin-	**cochin**eal (a brilliant scarlet dye)
shining, shiny	lucid-	stratum **lucid**um
	nitid-	**nitid**ous
silver	argent-	**argent**ic
	argyr-	**argyr**ophil
snow white	niv-	**niv**eous
sooty	fum-i	**fum**igant
	fuligin-	**fuligin**ous
spotted	macula-	**macula**te
straw-colored	stramin-	**stramin**eous
streaked	plagat-	**plagat**eous
tawny	fulv-	**fulv**ous
	cirrh-o	**cirrh**otic liver
violet	ianthin-	**ianth**ic, **ianthin**us
	iod-	**iod**ine
	viol-	**viol**escent
waxy	cera-	**cera**ceous

Color / Qualifier	Root Word / Prefix	As In
white	alb-	linea **alba**, **alb**umin
	calc-	**calc**areous (chalky)
	cand-	**cand**escence (gleaming)
	leuc-o	**leuc**ocyte
	etiol-	**etiol**ation
	ebrun-	**ebrun**eous (ivory white)
wine	oen-o	**oen**ophilic
yellow	citr-	**citr**eous (lemon-yellow)
	flav-o	**flav**inoids, **flav**ous
	lute-	**lute**ous (greenish-yellow)
	sulphur-	**sulphur**eous (pale-yellow)
	xanth-o	**xanth**ophyll

Size

Size	Root Word / Prefix	As In
dwarf	nan-o	**nan**ometre
	humil-	**humil**is (lowly)
	pumil-	**pumil**ate
equal	equ-	**equ**atorial plane
	is-o	**is**otonic
	pari-	**pari**pinnate
gigantic	coloss-	**coloss**al
	gigant-	**gigant**ism
	titan-	**titan**ic
	pelor-	**pelor**ic (monstrous)
large	grand-o	**grand**iose
	macr-o	**macr**oscopic
	magn-i	**magn**ification
	mega-	**mega**vitamin
	major	labia **major**a
	maxim-	**maxim**um
less than	mei-	**mei**osis
	sub- (below)	**sub**madibular
long	long-	**long**itudinal
	mec-o (height)	**Mec**optera (an order of insects with long, narrow wings)

Size	Root Word / Prefix	As In
medium	medi-	**medi**al section
short	brachy-	**brachy**dactylic
	brev-	**brev**irostrate
small	micr-o	**micr**oscopic
	mini-	**mini**ature
	minor	labia **minora**
stout	robus-	**robus**tus
tall	alti-	**alti**tude
	elat-	**elat**or
unequal	anis-o	**anis**ogamy

Shape and Texture

The suffixes -oid, -ode, -form, and -ate are commonly used to denote shape, form, and sometimes texture.

Shape / Texture	Root Word / Prefix	As In
angled, angular	angul-	**angul**ar
	gon-i	**gon**ion, dia**gon**al
arrowhead-like	hast-	**hast**ate
	sagitt-	**sagitt**ate
asymmetrical	oblique-	**oblique**
axe-like	dolab-	**dolab**riform
bag-like	utricul	**utricul**ate
bare	gymn-o	**gymn**ospermae
	nud-	**nud**ation
	psil-o	**Psil**ophyta
barrel-shaped	doli-	**doli**form
beak-like	corac-o	**corac**oid
	rostr-	**rostr**ate
bean-like	fab-	**fab**iform
bearded	barb-	**barb**ate

Shape / Texture	Root Word / Prefix	As In
bell-like	campan-	**campan**ulate
bent	ancyl-	**ancyl**oid
	arc-i	**arc**iform
	campt-	**campt**otropism
blunt, dull	ambly-	**ambly**opia
	obtus-	**obtus**e
	retus-	**retus**e
branched	ram-	**ram**ose
bristly	hispid-	**hispid**ate
	set-	**set**ae
broad and flat	lamin-	**lamin**ate
	lamell-	**lamell**iform
broom-like	corem-	**corem**oid
brush-like	penicill-	**penicill**iform, ***Penicillium***
	scop-	**scop**iform
bulging	mamill-	**mamill**ose
bumpy, nipple	papill-	**papill**ose
bushy	dum-	**dum**ose
chalky (limy)	calc-	**calc**areous
circular	orb-	**orb**icular, **orb**ate
	cycl-o	**cycl**oid
claw-like	chel-	**chel**iform
climbing	scand-	**scand**ent
club-like	clavat-	**clavat**e
	rhopal-	**rhopal**oid
closed	cleist-o	**cleist**othecium
cloth-like	pann-	**pann**ose
clustered	fasc-i	**fasc**iculate
	acin-	**acin**aceous
coiled	heli-	**heli**x, **heli**cal
comb-like	cten-o	**cten**oid
	pectin-	**pectin**ate

Shape / Texture	Root Word / Prefix	As In
cone, top-like	strobil- turb-	**strobili**form, **strobil**ate **turb**inate
convex	umbo-	**umbo**nate
creeping	rept-	**rep**ant, **rept**ile
crescent	lunul- menisc-	**lunul**ate, **lunul**ar **menisc**us, **menisc**oid
crested	crista-	**crista**
crooked	ankyl- scoli-	**ankyl**ostomiasis **scoli**osis
cross-like	cruci-	**cruci**ate, **cruci**form
crystal (prism)	prism-	**prism**atic
cup-shaped	caly- cyath- scyph-	**caly**coid **cyath**iform **scyph**iform
curled	crisp- bostrych-	**crisp**ate **bostrych**ate
curved	sinu-	**sinu**ose
curved back	recurv-	**recurv**ate
cylindrical	cylind- fistul-	**cylind**rical **fistul**ose
dense	(see thick)	
downy	papp- pubes-	**papp**ous **pubes**cent
dropping (nodding)	cernu-	**cernu**ous
dull	ambly-	**ambly**opia
dwarfish	nan-o	**nan**oid
ear-like	aur-	**aur**iculate
egg-like	oo-, ov-	ovoid, **ov**ate
elliptical	ellip-	**ellip**tical
entire (margin)	hol-o	**hol**ozoid
erect	corall-	**corall**oid
fan-like	flabelli-	**flabelli**form

Shape / Texture	Root Word / Prefix	As In
feather-like	pinn-	**pinn**ate
	plum-	**plum**ule
fiber-like	fibr-i	**fibr**illose, **fibr**ous
finger-like	digit-	**digit**ate
	dactyl-	**dactyl**ate
flask-shaped	lagen-	**lagen**iform
flat	plan-	***Plan**arium*
	platy-	**Platy**helminthes
flexible	camp-o	**camp**otrichia
forked	dich-	**dich**otomous
	furc-	bi**furc**ate
form, shape	-form	reni**form**
	morph-o	**morph**ology
	schemat-	**schemat**ic
fringed	lacin-	**lacin**iform
furrowed (grooved)	stria-	**stria**ted
	sulc-	**sulc**ate
gland	aden-o	**aden**oid
glassy	vitr-	**vitr**eous
glossy, varnished	verni-	**verni**cose
granular	sesam-	**sesam**oid
hairy	crin-	**crin**ose
	trich-o-	**trich**ophore, **trich**ome
	vill-	**vill**ose
hammer-shaped	mall-	**mall**eate
hand-like	palm-	**palm**iform, **palm**ate
hanging down	pend-	**pend**ulous
hard	dur-	**dur**a mater, in**dur**ation
	scler-	**scler**osis
heart-shaped	cord-	**cord**ate
heavy	gravid-	**gravid** (laden with young)
helmet-shaped	gale-	**gale**ate, **gale**iformis

Shape / Texture	Root Word / Prefix	As In
hollow	alve-	**alve**olar
	cav-i	**cav**ity
	coel-o	**coel**om
	glen-o	**glen**oid
honeycomb	fav-	**fav**eolate
hooked	adunc-	**adunc**ate
	ancyl-	**ancyl**oid
	gryph-	**gryph**oid
horn-like	cerat-o	*Ceratium*, **cerat**oid
	corn-	**corn**eous
humped	gibb-	**gibb**ous
inflated (puffy)	inflat-	**inflat**ed
irregular, gnawed	er-	**er**ose, **er**oded
keel-like	carin-	**carin**ate
kidney-like	nephr-o	**nephr**oid
	ren-	**ren**iform, **ren**al
knob-like	umbo-	**umbo**nate, **umbo**
knot-like projection	tyl-o	**tyl**oid
ladder-like	scal-	**scal**ariform
lance-like	lance-	**lance**olate
layered	strat-	**strat**ified
leaf	foli-o	**foli**ose
	frond-	**frond**iform
	phyll-o	**phyll**ode
leathery	cori-	**cori**aceous
lens-shaped	lent-i	**lent**icular, **lent**iform
linear (in rows)	line-	**line**ar
	ser-	**ser**ial sections
lipped	labi-	**labi**ate
lobed	lob-	**lob**ate, **lob**ule
	palm-	**palm**ate
long	elong-	**elong**ated
	long-i	**long**itudinal

Shape / Texture	Root Word / Prefix	As In
loose	lax-	**lax**ative
marked	notat-	**notate**
mealy	farin-	**farin**ose
membrane-like	hymen-	**hymen**oid
monstrous	terat-o	**terat**oid
moss-like	musc-	**musc**iform
mushroom-like	fung-	**fung**iform
naked, nude	nud-	**nude**
narrow	angusti- lin- sten-o	**angusti**form **lin**ear **sten**oid
navel-like	omphal-	**omphal**ode
needle-like	acicul-	**acicul**ate
network	reti- reticul-	**reti**form, **reti**na **reticul**ate-net
nipple-like	papill-	**papill**ate
node-like	nod-	**nod**ulate
oblique	oblique-	**oblique**
open	aper-i	**aper**ture
oval	ellip-	**ellip**soid
overlapping	imbric-	**imbric**ate
paired, twin	didym-	epi**didym**is, **didym**ous
palm-like	palm-	**palm**ate
pear-shaped	pyri-	**pyri**form
plank-like	tabl-	**tabl**oid
pointed	acu- mucr-	**acu**tate, **acu**leate **mucr**onate
pouched	sacc-	**sacc**ate
prostrate	procumb-	**procumb**ent
radiating	radi-	**radi**ate

Shape / Texture	Root Word / Prefix	As In
ragged	rhag-	**rhag**ose
reclining	decumb-	**decumb**ate
ribbon-like	taen-	**taen**ianus
ring-like	annul-	**annul**ar
rod	rhabd-	**rhabd**oid
rolled inward	invol-	**invol**ute
rolled back	revol-	**revol**ution
rough	scab-	**scab**rous
round	circ-	**circ**ular
	cocc-i	**cocc**oid
	cycl-	**cycl**e, **cycl**ic
	glob-	**glob**ular
	rot-	**rot**ate
round and flat	disc-, disk-	**disc**oid
S-shaped	sigm-	**sigm**oid
sac-like	sacc-	**sacc**ate
saliva-like	sial-	**sial**oid
saucer-shaped	pater-	**pater**iform
sausage-shaped	allant-o	**allant**oid
scaly	scurf-	**scurf**eous
	squam-	**squam**iform, **squam**ous
scalloped	cren-	**cren**ulate, **cren**ation
shaggy, hairy	dasy-	**dasy**phyllous
shape	(see form)	
sharp	acr-	**acr**id
	acu-	**acu**leate
	oxy-	**oxy**genate
shield-like	pelt-	**pelt**ate
ship-like	nav-i	**nav**icular
shrub-like	frutic-	**frutic**ose
sickle-shaped	falc-	**falc**iform
	drepan-	**drepan**iform

Shape / Texture	Root Word / Prefix	As In
sieve-like	cribr-	**cribr**iform
	ethm-o	**ethm**oid
silky	seri-	**seri**ceous
slanted	decliv-i	**decliv**inate
slender, thin	gracil-	**gracil**is muscle
	lept-	**lept**oid
	tenu-	**tenu**ate
slug-like	lima-	**lima**ciform
smooth	glabr-o	**glabr**ous
	laevi-	**laevi**gate
snake-like	oph-i	**oph**iolate
soft, flabby	malac-	**malac**ology
	mollusc-	**Mollusc**a (**Molluska**)
	flacc-	**flacc**id
spherical, round	cocc-i	**cocc**oid
	glob-o	**glob**ose
	rot-	**rot**ate
	spher-o	**spher**oid
spider-like	arachn-	**arachn**oid
spindle	fus-i	**fus**iform
spiny, thorny	acanth-	**acanth**ous
	echin-o	**echin**oid
	spin-	**spin**ose
spiral	spir-	**spir**ochete, **spir**al
	stromb-	**stromb**oid
split	dich-	**dich**otomous
	schist-	**schist**osome
	schiz-	**schiz**oid
spoon-like	spatul-	**spatul**ate
spreading, expanded	pat-	**pat**ent
square	cub-	**cub**iform
	quadr-	**quadr**ate
square-cut	trunc-	**trunc**ated

Shape / Texture	Root Word / Prefix	As In
stalk-like, stalked	pedic- styl-	**pedic**ulate **styl**oid
star-shaped	actin-o aster- stell-	**actin**omorphic **aster**oid **stell**ate
steep	anant- ardu-	**anant**oid **ardu**ous
stiff, upright	rig-	**rig**id
sticky	glutin- visc-	**glutin**ous **visc**ous
straight	euthy- orth-o rect-	**euthy**comous **orth**opedic **rect**ate, **rect**um
strap-like	ligul-	**ligul**ate
sword-shaped	ens- xiph-	**ens**iform **xiph**oid
tail-like (with a tail)	ur-o	**U**r**o**chordate
tapered (base to tip)	lance- subul-	**lance**olate **subul**ate
tapered (to a point)	acumin-	**acumin**ate
teardrop	lacr-i gutt-	**lacr**iform **gutt**ation
teeth-like	dent-, dens-	**dent**ate
tendril-like	capreol-	**capreol**ate
thick, dense	hadr- pachy- pycn-	**hadr**ate **pachy**derm **pycn**iform
thick, short	crass-	**crass**ulaceous
thin (thread-like)	fil- nemat-	**fil**iform **nemat**ode
thin (membrane-like)	scar-i	**scar**iform, **scar**ious
tongue-like	gloss-, glott-	**gloss**oid
top-like	(see cone)	

Shape / Texture	Root Word / Prefix	As In
torn	rhag-	**rhag**oid
trapezoid	trapez-	**trapez**ius
tree-like	arbor-	**arbor**escent
	dendr-	**dendr**oid, **dendr**ite
triangular	delt-o	**delt**oid
	trigon-	**trigon**ous
tufted	loph-	**loph**oid
turnip-shaped	nap-	**nap**iform
twisted	contort-	**contort**ed
	plect-o	**plect**oid
	strept-o	**strept**ococcoid
	strobil-	**strobil**us
	tort-	**tort**uous
U-shaped	hy-	**hy**oid
uneven, unequal	anis-o	**anis**ogamy
velvety	velutin-	**velutin**ate, **velutin**ous
wavy	kym-	**kym**oid
	undulat-	**undulat**ed
waxy	glauc-o	**glauc**ate, **glauc**ous
web-like	tela-	**tela**te
wedge-like	cune-	**cune**iform
	sphen-o	**sphen**oid
whip-like	flagell-	**flagell**iform
whorled	verticill-	**verticill**ate
wide, broad	eury-	**eury**baric
	lat-	**lat**isimus
	platy-	**platy**sma
wing-like	al-	**al**iform
	penn-	**penn**ate
	pinn-	**pinn**ate
	pteryg-o	**pteryg**oid
withered	marces-	**marces**cent
wood-like	lign-	**lign**eous

Shape / Texture	Root Word / Prefix	As In
woolly	lan-	**lan**ate
	mallo-	**mallo**id
worm-like	helminth-	**helminth**oid
	verm-	**verm**iform
wrinkled	rug-	**rug**ose
Y-shaped	hy-	**hy**oid

Direction and Position

Direction / Position	Root Word / Prefix	As In
above, over	hyper-	**hyper**tonic
	super-	**super**ficial
	supra-	**supra**maxilla
across, through	dia-	**dia**lysis
	trans-	**trans**location
after, next to	meta-	**meta**phase
against, opposite, toward	ad-	**ad**nate
	af-	**af**ferent
	ag-	**ag**grade
	anti-	**anti**clinal
	contra-	**contra**ceptive
	ob-	**ob**tuse
apart	dia-	**dia**lysis
	dis-	**dis**junction
apex, top	acr-o	**acr**omial
	apic-	**apic**al meristem
around, on both sides of	ambi-	**ambi**tal
	circum-	**circum**orbital
	peri-	**peri**clinal
away from, down	ab-	**ab**ductor, **ab**axial
	ap-o	**ap**ositional
backward, behind	opisth-o	**opisth**osoma
	retro-	**retro**gression

Direction / Position	Root Word / Prefix	As In
before, in front of	anter-	**anter**ior
	pre-	**pre**zygotic
	pro-	**pro**phase
behind	post-	**post**erior, **post**zygotic
below, under	hyp-o	**hypo**tonic
	infra-	**infra**orbital
	sub-	**sub**mandibular
	sus-	**sus**pended
beside, near	para-	**para**physis
between, among	inter-	**inter**phase, **inter**stial
	meta-	**meta**phase
beyond	ultra-	**ultra**structure
down	cata-	**cata**bolic
	decid-	**decid**uous
far	tele-	**tele**scopic
first	arche-	**arche**nteron
	prim-	**prim**ate, **prim**ary consumer
	prot-o	**proto**derm
from	ex-o	**exo**crine
inner, within, inside	endo-	**endo**dermis
	ento-	**ento**cyclic
middle	medi-	**medi**al
	mes-o	**meso**derm
	mid-	**mid**-rib
near	para-	**para**thyroid
	proxim-	**proxim**al
on, on top of	epi-	**epi**dermis
outside, out of	ecto-	**ecto**derm
	ex-o	**exo**skeleton
	extern-	**extern**al
	extra-	**extra**cellular
second	deuter-o	**deutero**stome
side	later-	**later**al
	pleur-o	**pleur**al ganglia
through	per-	**per**meable, **per**foration

together, with	syn- sym-	**syn**apse **sym**biotic

Number and Quantity

Number / Quantity	Root Word / Prefix	As In
one	mono- uni-	**mono**clonal **uni**sexual
two	bi- di-	**bi**sexual **di**ssection
three	tri- terti	**tri**ceps **terti**ary
four	quadr- tetra- quater-(fourth)	**quadr**aceps **tetra**ploid **quater**nary
five	pent-a quinque- quint-(fifth)	**pent**aradial **quinque**fid **quint**uplet
six	hex- sext-	**hex**agonal **sext**uplets
seven	hepta- septem-	**hepta**merous **Septem**ber
eight	octo- octav- (eight)	**octo**pus **octave**
nine	ennea- novem-	**ennea**ndrous **Novem**ber
ten	deca- deci- (tenth)	**deca**de **deci**metre
eleven	hendeca-	**hendeca**gon
twelve	dodeca-	**dodeca**gon
all	pan- pant-o toti-	**Pan**gea **pant**ology **toti**potent
countless	myri-o	**myria**sporous
double	didym dipl-o duplex-	epi**didym**is **dipl**oid **dupl**ication

Number / Quantity	Root Word / Prefix	As In
double, of two kinds	amphi-	**amphi**bian
equal	equ-	**equ**ation
	is-o	**is**omorphic
	par-i	**par**ipinnate
few	olig-o	**olig**oglia
	pauc-i	**pauc**ity
first	prim-	**prim**ary
many, numerous	multi-	**multi**fidus
	poly-	**poly**ploidy
	plei-o	**plei**otrophic
more	plio-	**Plio**cene
	plur-	**plur**ilocular
one and a half	sesqui-	**sesqui**oxide
one half	hemi-	**hemi**sphere
	semi-	**semi**permeable
one hundred	hecto-	**hecto**gram
one thousand	kilo-	**kilo**gram
several, most	plur-	**plur**ilocular
	plei-o	**plei**otropism
single	hapl-o	**hapl**oid
unequal	anis-o	**anis**ogamy
whole	holo-	**holo**zoic
	integ-	**integ**ral

Exponential Notation

Writing very small and very large numbers is awkward, and trying to actually work with such numbers can be very cumbersome. The use of exponential notation is much easier and the chance of error is greatly reduced. This simplified method involves expressing numbers in "powers of tens" and by Greek and Latin prefixes. The following prefixes can be used with any SI (System International) unit (note that the combining vowel is part of the exponential prefix):

Exponential Form	Prefix	Symbol	As In
10^{18}	exa-	E	**exa**metre
10^{15}	petra-	P	**petra**metre
10^{12}	tera-	T	**tera**metre
10^{9}	giga-	G	**giga**joule
10^{6}	mega-	M	**mega**pascal
10^{3}	kilo-	k	**kilo**gram
10^{2}	hecto-	h	**hecto**gram
10^{1}	deca-	da	**deca**metre
10^{0}			
10^{-1}	deci-	d	**deci**metre
10^{-2}	centi-	c	**centi**metre
10^{-3}	milli-	m	**milli**metre
10^{-6}	micro-	μ	**micro**gram
10^{-9}	nano-	n	**nano**seconds
10^{-12}	pico-	p	**pico**gram
10^{-15}	femto-	f	**femto**metre
10^{-18}	atto-	a	**atto**metre

Glossary of Common Biological Terms

A

ABA—see abscisic acid

abaxial (*ab-*, away from, + *-ax-*, center line, axis, + *-ial*, pertaining to): facing away from the axis, usually in reference to the stem of a plant. Opposite of adaxial.

abdomen (*abdomen*, belly): (1) in vertebrates, the region of the body between the diaphragm and the pelvis containing the viscera. (2) in arthropods, the posterior region of the body containing reproductive and digestive organs.

abductor (*ab-*, away, + *-duc-*, to lead, + *-or*, state of): a muscle that moves a structure away from the midline (axis) of the body; such as raising the arm laterally. Opposite of adductor.

abiogenesis (*a-*, without, + *-bi-*, life, + *-gen-*, origin, + *-sis*, the process of): a presently accepted explanation for the origin of life from nonliving (inorganic) materials.

abiotic (*a-*, without, + *-bi-*, life, + *-tic*, pertaining to the process of): nonliving factors or agents that directly affect life; such as water, wind, temperature, gases, etc. Opposite of biotic.

aboral (*ab-*, away, + *-or-*, mouth, + *-al*, pertaining to): a region of an organism opposite the oral (mouth) region. Opposite of oral.

abscisic acid (*abscis-*, cut off, + *-ic*, pertaining to, + acid): a plant hormone involved with dormancy, abscission, and water stress; abbreviated ABA.

abscission layer (zone) (*abscis-*, cut off, + *-ion*, process of, + layer): a layer of cells at the base of an organ, such as a fruit, leaf, or flower, which becomes weak, causing the loss of the organ.

absorption spectrum: a measure of the amount of light energy a pigment can absorb at a specific wavelength, usually in reference to photosynthesis.

abyssal zone (*abys-*, bottomless, + *-al,* pertaining to): a region of the ocean floor where temperatures are cold, pressures intense, and where light is totally absent.

accessory bud (shoot) (*access-*, supplementary, + *-ory,* related to, + bud): a bud usually adjacent to and smaller than the primary apical bud.

accessory cells (*access-*, supplementary, + *-ory,* related to, + *cell)*—**see subsidiary cells**

accessory fruit (*access-*, supplementary, + *-ory,* related to, + bud): a fruit derived from tissues other than the ovary, such as an apple (a pome), which is derived from the hypanthium. See pome.

accessory pigment (*access-*, supplementary, + *-ory,* related to, + bud): a photosynthetic pigment other than chlorophyll ***a*** that is involved in absorbing light energy; such as the carotenoids. See carotenoid.

acclimation (*a-*, to, + *-clim-*, region, + *-ation,* process of): one or more physiological adjustments made in response to a change in the environment.

accretion (*accres-*, to increase, + *-tion,* the process of): growth by the adding on of new material or cells to the external surface; such as the growth of corals.

acetabulum (*acetabul-*, a vinegar cup, + *-um,* structure): a socket in the pelvic bone to which the femur attaches.

acetyl-coenzyme A (acetyl CoA): an important intermediate compound linking glycolysis to the Krebs cycle; made of an acetyl group covalently bonded to coenzyme A.

acetylcholine: a neurotransmitter responsible for the movement of nerve impulses across synapses; the acetyl ester of choline.

achene (*a-*, without, + *-chen-*, a gap): a small, dry, nonsplitting fruit containing one seed; such as a sunflower "seed" that technically is the entire fruit.

acid (*acid-*, sour): any substance that releases or donates hydrogen ions (protons) in solution producing a pH of less than 7 and having a sour taste. Opposite of a base.

acid rain: rainwater with a pH less than 5.6; usually caused by the presence of sulfur oxides and/or nitrogen oxides in the atmosphere that react with water to form sulfuric acid and/or nitric acid.

acid-growth hypothesis: the proposed mechanism by which auxin affects plant growth by inducing plant cells to elongate by loosening the molecular structure of cell walls.

acidophil (*acid-*, sour, + *-phil,* to love): a white blood cell whose cytoplasmic granules readily stain red with acidic dyes such as eosin. Also referred to as eosinophil.

acoelomate (*a-*, without, + *-coel-*, cavity, + *-ate,* to form): a solid-bodied animal lacking a true body cavity (coelom) between the gut and body wall; such as a flatworm. Opposite of eucoelomate. See coelom.

acontium, ***pl.*** **acontia** (*acont-*, dart, + *-ium,* region): a relatively long thread-like strand found in sea anemones studded with stinging cells (nematocysts) and capable of being discharged through the mouth of the organism; used for capturing food.

Acquired Immune Deficiency Syndrome (AIDS): an extremely contagious disease caused by a retrovirus (Human Immunodeficiency Virus) that infects and destroys vital components of the immune system; no known cure to date.

acropetalous (*acr-*, tip, + *-petal-*, a flower leaf, + *-ous,* pertaining to): a pattern of development in plants starting at the base and proceeding toward the tip of an organ; usually in reference to flowering. Opposite of basipetal.

acrosome (*acr-*, tip, + *-som-*, body): a cap-like structure covering the tip of animal sperm cells containing enzymes that aid the sperm cell to penetrate and fertilize the egg; also referred to as apical body.

actin (*actin-*, ray): a globular protein that, with myosin, is responsible for muscle contraction; the major component of thin filaments in myofibrils. Actin is also a major component of the cytoskeleton. See myosin.

actinomorphic (*actin-*, ray, + *-morph-*, form, + *-ic,* pertaining to)—**see radial symmetry**

actinostele (*actin-*, ray, + *-stel-*, a pillar): a type of protostele in which the vascular tissue is arranged in radiating arms separated by parenchyma. See protostele.

action potential: the polarity of the membrane of muscle cells and nerve cells during contraction or during the conduction of a nerve impulse. The inside of the cell is more positive than the outside; also referred to as nerve impulse. Opposite of resting potential.

action spectrum: a measure of the amount of energy needed to trigger a specific chemical reaction, such as photosynthesis; usually measured in nanometres.

activation energy: the amount of energy needed to initiate a chemical reaction; usually measured in kilocalories.

active site: the region(s) on the surface of an enzyme that binds the substrate molecule(s) to the enzyme and catalyzes its reaction with another substrate.

active transport: a transport mechanism requiring energy and a carrier molecule that moves substances across a cell membrane against a concentration gradient (from a low concentration to a high concentration). This mechanism is analogous to a pump requiring energy to move water against the force of gravity.

adaptation (*adapt-*, to fit, + *-ation,* the process of): (1) the ability of an organism to adjust to its enviroment for it to survive and reproduce, either physiologically or structurally. (2) a decrease in the response of a nerve receptor exposed to repeated or prolonged stimulation.

adaptive radiation (*adapt-*, to fit, + *-ive,* tending to, + *radi-*, a spoke of a wheel, a ray, + *-ation,* the process of): the evolution of several new specialized species from one unspecialized ancestor. Each new species is adapted to fill a specific niche; also referred to as divergent evolution, speciation, or macroevolution.

adaxial (*ad-*, toward, + *-ax-*, center line, axis, + *-al,* pertaining to): facing toward the axis; usually in reference to the stem of a plant. Opposite of abaxial.

adductor (*ad-*, toward, + *-duc-*, to lead, + *-or*, state of): a muscle that moves a structure toward the midline (axis) of the body; such as lowering the arm laterally. Opposite of abductor.

adenine (*aden-*, gland, + *-ine*, having the character of): one of four nitrogenous nucleotide bases that is a component of nucleic acids (DNA and RNA); complimentary to thymine; also an important component of energy transfer molecules such as ATP, ADP, and AMP.

adenoid (*aden-*, gland, + *-oid*, resembling): lymphatic tissue located in the nasopharynx; also in reference to anything gland-like.

adenosine (mono-, di-, tri-) phosphate: a nucleotide containing adenine, ribose, sugar, and one (AMP), two (ADP), or three (ATP) phosphate groups. All three are essential in the storage and transfer of energy in biological systems.

adenosine diphosphate (ADP)—see adenosine (mono-, di-, tri-) phosphate

adenosine monophosphate (AMP)—see adenosine (mono-, di-, tri-) phosphate

adenosine triphosphate (ATP)—see adenosine (mono-, di-, tri-) phosphate

adhesion (*adher-*, to stick to, + *-ion*, the process of): the attraction and binding of unlike molecules; such as water to glass.

adipose (*adip-*, fat, + *-ose*, resembling): fatty tissue, or the fat itself.

adnation (*adn-*, to grow on or to, + *-ation*, the process of): the fusion of unlike parts; such as stamens and petals.

adrenal glands (*ad-*, toward, + *-ren*, kidney, + *-al*, pertaining to): paired endocrine glands located just above mammalian kidneys; involved in controlling water/salt balance and physiological reactions to stress.

adrenaline—see epinephrine

adsorption (*ad-*, toward, + *-sorb-*, to suck in, + *-tion*, the process of): the adhesion of molecules, either liquid, gas, or solid, to a solid; such as the adsorption of water to seeds.

adventitia (*adventit-*, foreign, + *-ia*, condition of): a type of loose connective tissue that forms a thin membrane-like covering over some vertebrate organs; also referred to as fascia.

adventitious (*adventit-*, foreign, + *-ous*, pertaining to): a reference to structures developing from an unusual location; such as roots growing from leaves.

aerate (*aer-*, air, + *-ate*, to form): to provide a supply of gas; such as oxygen to roots.

aerenchyma (*aer-*, air, + *-en-*, in, + *-chym-*, infusion): plant tissue containing a large number of air spaces.

aerobic (*aer-*, air + *-bi-*, life + *-ic*, pertaining to): any metabolic process that requires molecular oxygen (O_2); such as aerobic respiration. Opposite of anaerobic.

afferent (*af-*, toward, + *-fer-*, to carry, + *-ent*, performing the action of): pertaining to structures carrying something toward an organ or region; such as blood vessels and nerves. Opposite of efferent.

afterbirth—see placenta

agar (*agar*, seaweed): a polysaccharide extracted from the cell walls of certain red algae and commercially marketed as a solidifying agent used in laboratory media.

agglutination (*agglutin-*, glued together, + *-ation*, the process of): the visible clumping together of particles or cells in solution, as when different blood types within the same species are mixed.

aggregate fruit (*ad-*, to, + *-grega-*, to collect together, + *-ate*, characterized by having): a type of fruit derived from the fusion of several separate carpels from one flower; such as a raspberry.

agonist (*agon-*, to fight, + *-ist*, that which does): a muscle that produces the desired action when contracting. Opposite of antagonist.

agranulocyte (nongranular leukocyte) (*a-*, without, + *-gran-*, granular, + *-cyte*, cell): one of two classes of white blood cells that does not contain distinct cytoplasmic granules. T and B lymphocytes and monocytes are agranulocytes. The other class is granulocytes.

AIDS—see Acquired Immune Deficiency Syndrome

alate (*al-*, wing, + *-ate*, characterized by having): winged; often associated with wing-like projections on certain bones.

albinism (*albin-*, white, + *-ism*, the process of): a genetic disorder that prevents the production of melanin (a pigment in skin, hair, and eyes).

aldehydes (a contraction of alcohol and dehydrogenated): a common group of organic compounds with a carbonyl group located at the end of a carbon chain.

aldosterone (*aldo-*, to nourish, + *-ster-*, hard, solid): a steroid hormone involved in regulating sodium/potassium levels in the blood; produced by the adrenal cortex.

aleurone layer (*aleuron-*, flour, + layer): the outermost layer of endosperm tissue containing hydrolytic enzymes associated with digesting the endosperm tissue; typical of monocot seeds.

alga, *pl.* algae (*alg-*, seaweed): a nontaxonomic grouping of simple, mostly aquatic organisms; essential as primary producers.

algin (*alg-*, seaweed, + *-in*, chemical substance): a polysaccharide extracted from certain brown algae; commercially marketed as an emulsifying agent.

alimentary canal (*aliment-*, food, + *-ary*, apparatus, + canal): the tube that extends from the mouth to the anus and functions as a digestive tract.

alkaline (alkali) (*alkal-*, basic, + *-ine*, having the character of): any substance that releases or donates hydroxide ions (OH^-) in solution producing a pH greater than 7; also referred to as a base. Opposite of acid.

alkaloids (*alkal-*, basic, + *-oid*, resembling): naturally occurring nitrogenous compounds that often have toxic and hallucinogenic effects on animals; such as caffeine, nicotine, strychnine, and morphine.

allantois (*allant-*, sausage, + *-oid*, resembling): one of four membranes associated with the embryos of birds, reptiles, and mammals and functions to store nitrogenous wastes excreted from the embryo; a source of blood vessels to and from the placenta in mammals; located between the chorion and amnion layers.

alleles (*allel-,* of one another): alternate forms of a single gene coding for the same trait; situated at the same position (locus) on homologous chromosomes. For example, a gene for height may have an allele for tallness and an allele for shortness.

allergy (*all-,* other, + *-erg-,* work, activity): a heightened sensitivity or susceptibility to a foreign compound (often a protein).

allopatric speciation (*all-,* other, + *-patr-,* fatherland, + *speci-,* a kind, + *-ation,* process of): the evolution of new species resulting from the ancestral population becoming segregated geographically into two or more groups. Opposite of sympatric speciation.

allosteric enzyme (*all-,* other, + *-ster-,* hard, solid, + *-ic,* pertaining to, + enzyme): an enzyme that alters its shape and catalytic activity due to the binding of a smaller allosteric effector "control" molecule.

alpha cells (*alpha,* first, + cell): cells in the islets of Langerhans (pancreas) that produce and secrete glucagon. See glucagon.

alpha (α) helix (*alpha,* first, + *helix,* spiral): a secondary structure of proteins formed when a polypeptide chain regularly turns about itself to form a "spiral staircase" shape. See secondary protein structure.

alternating cleavage—see spiral cleavage

alternation of generations: the alternating between a haploid (n) gamete-producing plant (gametophyte) and a diploid (2n) spore-producing plant (sporophyte); associated with sexual reproduction.

altrical (*altric-,* nourishers, + *-al,* pertaining to): in reference to animals with young who hatch in a totally dependent, immature state; usually in reference to birds.

alveolus, *pl.* alveoli (*alveol-,* small cavity, + *-us,* thing): (1) a small hollow dead-end cavity in lungs in which gas exchange occurs; also referred to as an air sac. (2) a tooth socket. (3) a secretory unit of an exocrine gland, such as a milk-secreting sac in mammary glands.

ameboid (amoeboid) movement (*amoeb-,* change, + *-oid,* resembling): a type of locomotion characterized by the formation of pseudopodia and the slow oozing of cytoplasm into the pseudopodia. Characteristic of macrophages and *Amoeba.*

amino acids (*Ammon,* the temple of Zeus where ammonium salts were prepared from dung): organic compounds containing an amino group ($-NH_2$) and a carboxyl group (–COOH) both bonded to the same carbon atom; often referred to as the "building blocks" (monomers) of proteins.

amino group: a common functional group consisting of a nitrogen atom bonded to two hydrogen atoms; often acting as a base in solution.

aminoacyl-tRNA: an organic compound consisting of an amino acid linked covalently to tRNA.

ammonification (*Ammon,* the temple of Zeus where ammonium salts were prepared from dung, + *-fic-,* to make, + *-ation,* the process of): the process of breaking down proteins and releasing nitrogen in the form of ammonia (NH_3) or ammonium (NH_4^+) back into the soil or atmosphere; performed by certain fungi and bacteria.

amniocentesis (*amni-*, membrane around the fetus, + *-cente-*, puncture, + *-sis*, process of): a procedure involving the removal of some amniotic fluid from the womb of a pregnant animal with a syringe to detect genetic disorders within fetal cells present in the fluid.

amnion (*amni-*, membrane around the fetus): the inner-most of four membranes around the embryos of mammals, birds, and reptiles that forms a fluid-filled sac around the embryo and protects the embryo.

amniote (*amni-*, membrane around the fetus): an animal whose embryos are surrounded by an amnion; reptiles, birds, and mammals. See amnion.

AMP—see adenosine monophosphate

amphipathic molecule (*amphi-*, on both sides of, + *-path-*, suffering, + *-ic*, pertaining to): a molecule having one part soluble in water (polar and hydrophilic) and another part insoluble in water (nonpolar and hydrophobic); such as a phospholipid or glycolipid molecule.

amphiphloic (*amphi-*, on both sides of, + *-phlo-*, bark, + *-ic*, pertaining to): pertaining to an arrangement of vascular tissue with phloem on both sides of the xylem.

ampulla (*ampull-*, flask): (1) a small flask-like vesicle at the base of a semicircular canal in the inner ear. (2) a pressurized bulb-like vesicle above each tube foot in the water-vascular system of echinoderms.

amylase (*amyl-*, starch, + *-ase*, enzyme): an enzyme found in animals that catalyzes the breakdown of starch into maltose; found in saliva and pancreatic juice.

amyloplast (*amyl-*, starch, + *-plast*, membrane): a colorless plastid used for storing starch; also referred to as a leucoplast. See plastid.

amylose (*amyl-*, starch, + *-ose*, resembling): a linear polymer of glucose; a relatively simple type of starch.

anabolic steroid (*ana-*, anew, + *-bol-*, to put, + *-ic*, pertaining to + steroid)**—see steroid**

anabolism (*ana-*, anew, + *-bol-*, to put, + *-ism*, the process of): the sum of all chemical reactions in which more complex molecules are formed from simpler ones; any synthesis reaction requiring energy, such as protein synthesis. Opposite of catabolism.

anaerobic (*an-*, without, + *-aer-*, air, + *-bi-*, life, + *-ic*, pertaining to): any metabolic process that does not require molecular oxygen (O_2); such as anaerobic respiration. Opposite of aerobic.

anagenesis (*ana-*, anew, + *-gen-*, origin, + *-sis*, process of): a form of evolution in which an entire population is, over time, changed into a new species; also referred to as phyletic evolution.

analogy (analogous structure) (*analog-*, ratio, + *-y*, state of): usually in reference to an anatomical structure that is similar in function or appearance to another but not in origin or development; such as the fins of a fish and the flippers of a whale. The independent evolution of similar structures or life forms as a result of convergent evolution.

anaphase (*ana-*, anew, + *-phase,* a stage): (1) the third stage of mitosis or of meiosis II during which the two sets of chromatids separate, forming chromosomes, and move to opposite poles of the cell. (2) the stage in meiosis I during which homologous chromosomes separate and move to opposite poles of the cell. See cell cycle.

anaphylaxsis (anaphylactic shock) (*ana-*, anew, + *-phylac-*, protection, + *-sis,* process of): a severe and usually sudden allergic reaction following exposure to a foreign substance (allergen); often life-threatening since blood flow to the brain and heart is reduced due to blood vessel constriction.

anastomosis (*ana-*, anew, + *-stom-*, mouth, + *-sis,* process of): the joining of two or more structures; such as blood vessels, nerves, or muscle fibers.

anatomy (*anat-*, to cut up, + *-my-*, muscle): the study of the internal and external structure of organisms.

androecium (*andr-*, male, + *-eci-*, dwelling, + *-um,* structure): the male part of a flower; a collective term for all the stamens. Opposite of gynoecium.

androgens (*andr-*, male, + *-gen-*, origin): male sex hormones that maintain secondary male sex characteristics; such as testosterone. Opposite of estrogens.

aneuploid (*an-*, without, + *-eu-*, true, + *-ploid,* multiple of): the loss or gain of one or more chromosomes; such as trisomy 21 (Down's syndrome), in which there is one extra twenty-first chromosome. See Down's syndrome.

angiosperm (*angi-*, enclosed, + *-sperm,* seed): a plant whose seeds are produced in and protected by a fruit (mature ovary). Also called a flowering plant or Anthophyte.

Angstrom (named after the Swedish physicist, A. J. Angstrom): a unit of length equal to 10 nanometres; symbolized, Å. The Angstrom is an obsolete unit and has been replaced by the nanometre.

anhydrase (*an-*, without, + *-hydr-*, water, + *-ase,* enzyme): an enzyme that removes water from a compound; such as carbonic anhydrase that catalyzes the conversion of carbonic acid into water and carbon dioxide.

animal starch—see glycogen

anion (*ana-*, up, + *-ion,* to wonder): a negatively charged ion produced by the addition of one or more electrons; such as Cl^-. Opposite of cation.

anisogamy (*anis-*, unequal, + *-gam-*, union, + *-y,* state of): a type of sexual reproduction in which both gametes are motile but unequal in size; typically a larger female and a smaller male. Also known as heterogamy.

annual (*annu-*, year, + *-al,* pertaining to): a plant that completes its reproductive cycle in one growing season.

annual ring (*annu-*, year, + *-al,* pertaining to, + ring): the layer of wood (secondary xylem) usually produced during one growing season.

annulus (*annul-*, ring, + *-us,* thing): (1) any ring-like structure. (2) a ring of cells in the sporangium of ferns. (3) a ring of cells encircling the opening of a moss capsule. (4) the remnants of the vellum remaining on the stipe of certain gill fungi.

antagonist (*anti-*, against, + *-agon-*, to fight, + *-ist,* that which does): a muscle that opposes the action of another muscle. See agonist.

antenna, ***pl.*** **antennae** (*antenna-,* sail yard): a paired sensory organ on the head of many arthropods that acts primarily as a chemoreceptor.

antennules (*antenna-,* sail yard, + *-ule,* little): paired sensory organs on the head of many arthropods that are smaller than antennae and that act primarily as chemoreceptors.

anterior (*anter-,* before, in front of, + *-or,* state of): the front- or head-end of an organism. Opposite of posterior.

anther (*anth-,* flower): the portion of a stamen that produces pollen; also referred to as a pollen sac.

antheridium (*anth-,* flower, + *-idium,* little): a multicellular, sperm-producing gametangium found in primitive vascular plants; such as mosses.

anthesis (*anth-,* flower, + *-sis,* the process of): the time required for flowering to occur.

anthocyanin (*anth-,* flower, + *-cyan-,* blue, + *-in,* chemical substance): a water-soluble red to blue pigment occuring in cell sap; particularly in the cells of flower petals.

anthropoid (*anthrop-,* man, + *-oid,* resembling): man-like; usually in reference to higher primates including monkeys, apes, and man.

antibiotic (*anti-,* against, + *-bi-,* life + *-tic,* pertaining to the process of): an organic compound produced by one organism (usually bacteria) that destroys or inhibits another organism's growth and development; often used in the treatment of infectious diseases.

antibody (*anti-,* against, + body): a complex globular protein (immunoglobulin) produced in response to the presence of a foreign substance (antigen) in the body and having the ability to react against the antigen.

anticlinal (*anti-,* against, + *-clin-,* to lean, + *-al,* pertaining to): a plane of cell division at right angles to a surface; as in an apical meristem. Opposite of periclinal.

anticodon (*anti-,* against, + *-code-,* writing tablet, + *-on,* particle): a triplet of nucleotides in tRNA that, in the process of protein synthesis, binds to a complimentary codon in mRNA.

antidiuretic hormone (ADH) (*anti-,* against, + *-dia-,* through, + *-ur-,* urine, + *-tic,* pertaining to the process of, + *hormone,* to excite): a hormone that induces the reabsorption of water by the kidneys and thereby controlling the amount of water excreted in the urine; also referred to as vasopressin.

antigen (*anti-,* against, + *-gen-,* origin): any substance capable of stimulating an immune response (production of an antibody); usually a foreign carbohydrate or protein.

antipodals (*anti-,* against, + *-pod-,* foot + *-al,* pertaining to): three cells located at the end opposite to the micropyle in angiosperm embryo sacs.

antler (*anter-,* before, in front of, + *-ler,* eye): a bony outgrowth from the head of such animals as moose, deer, and elk.

aorta (from *aeirein,* to raise up): the largest artery in the vertebrate body carrying oxygenated blood from the heart to the body.

aperture (*apertur-,* an opening): any opening, such as the opening on a snail shell.

apical body—see acrosome

apical dominance (*apic-*, tip, + *-al,* pertaining to, + *domin-*, ruling, + *-ance,* state of): a pattern of growth associated with the suppression of lateral bud growth by the apical bud, resulting in a tall (rather than bushy) plant.

apical meristem (*apic-*, tip, + *-al,* pertaining to, + *merist-*, divisible): a group of actively dividing embryonic cells at the tips of roots and shoots.

aplanospore (*aplan-*, nonmotile, + *-spor-*, spore): a nonmotile spore found in some algae and fungi.

apomixis (*ap-*, away from, + *-mix-*, intercourse, + *-sis,* the process of): the development of fruit and seeds without sexual reproduction, as in dandelions. Also referred to as parthenogenesis.

apomorphic (*ap-*, away from, + *-morph-*, shape, form, + *-ic,* pertaining to): usually in reference to the evolution of phenotypic traits in a lineage after it has branched off from a phylogentic tree.

apophysis (*ap-*, away from, + *-phy-*, to grow, + *-sis,* the process of): any outgrowth or projection, such as the bony outgrowths on vertebrae.

apoplast transport (*ap-*, away from, + *-plast,* membrane, + transport): the movement of materials via the cell walls and intercellular spaces; most obvious in plant root cells. Opposite of symplast transport.

apopyle (*ap-*, away from, + *-pyl-*, gate): the primary opening of the radial canal into the central cavity (spongocoel) of sponges.

apothecium (*apothec-*, storehouse, + *-ium,* region): an open cup-shaped ascocarp. See ascocarp.

appendage (*append-*, an addition, + *-age,* collection of): any structure attached to a larger structure; usually in reference to organs of locomotion, such as legs or fins.

appendix—see vermiform appendix

applied research: research that has an obvious and direct benefit to mankind. Most types of medical research are applied.

appositional growth (*appon-*, to put near to, + *-tion,* process of, + -al, pertaining to): the growth of tissue by the adding on of successive layers of matrix; such as the growth of bone. Opposite of interstitial growth.

aqueous (*aqua-*, water, + *-ous,* full of): a solution in which water is the primary solvent.

arboreal (*arbor-*, tree-like, + *-eal,* pertaining to): tree dwelling.

arborescent (*arbor-*, tree-like, + *-esc-*, to be somewhat, + *-ent,* having the quality of): resembling a tree in growth and/or in appearance, such as primitive nonwoody plants or the branching of certain neurons.

archegonium (*arch-*, primitive, + *-gon-*, reproduction, + *-ium,* region): a multicellular, egg-producing gametangium found in primitive vascular plants; such as mosses.

archenteron (*arch-*, primitive, + *-enteron-*, gut): the primary cavity of the gastrula stage of embryonic development that is lined with endoderm and represents the future digestive tract.

areolar (*areol-,* a small space, + *-ar,* pertaining to): (1) a relatively small space or area within tissue; such as those found between the fibers in loose connective tissue. (2) a colored ring around the nipple.

arteriole (*arter-,* artery, + *-ole,* little): a small artery that supplies blood to capillaries. Opposite of venule.

artery (*arter-,* artery): a thick-walled vessel that carries oxygenated blood away from the heart and toward tissues. Opposite of vein.

articulation (*articul-,* joint, + *-ation,* the process of): the joining of two or more structures, such as bones.

artificial selection: the selective breeding of domestic animals and plants for the purpose of obtaining offspring with certain desired traits.

ascocarp (*asc-,* cup, + *-carp-,* fruit): a reproductive structure lined with spore-producing cells called asci. Three distinct types exist in the Ascomycetes; see apothecium, perithecium, and cleistothecium.

ascorbic acid (*a-,* without, + *-scorb-,* scurvy, + *-ic,* pertaining to)—**see vitamin C**

ascospore (*asc-,* cup, + *-spor-,* spore): a spore produced within an ascus.

aseptate (*a-,* without, + *-sept-,* wall, + *-ate,* characterized by having): multinucleated algal or fungal filaments lacking cross walls; also called nonseptate.

asexual reproduction (*a-,* without, + sex, + *-al,* pertaining to, + reproduction): any reproductive process not involving the union of sperm and egg, resulting in offspring genetically identical to the parent; such as budding or binary fission. Opposite of sexual reproduction. Also referred to as vegetative reproduction.

assimilation (*assimil-,* to bring into conformity, + *-ation,* the process of): the formation of complex organic compounds from smaller ones that have been absorbed by the body. See anabolism.

aster (*astr-,* star): short microtubules radiating from the ends of the spindle apparatus in dividing animal cells.

asthma (*asthma-,* panting): a respiratory condition characterized by the narrowing of airways causing difficult breathing; often associated with being exposed to allergens such as pollen.

astrocyte (*astr-,* star, + *-cyte-,* cell): a cell with many cytoplasmic processes (star-like) that provides nutrients and insulation for neurons in the central nervous system; one of three types of neuroglial cells. See neuroglia cells.

asymmetry (*a-,* without, + *-symmet-,* measured together): a body plan without symmetry, as in some sponges. May be referred to as irregular symmetry. The opposite of symmetry.

atactostele (*atact-,* out of order, + *-stel-,* a pillar): a scattered arrangement of vascular tissue when viewed in cross section; common in monocot stems.

atherosclerosis (*ather-,* fatty plaque, + *-scler-,* hard, + *-sis,* process of): a cardiovascular disease characterized by deposits of fatty material in the walls of arteries, narrowing their diameter, leading to reduced blood flow and stroke.

atom (*atom-,* indivisible): the smallest unit into which a substance can be divided and still retain the chemical properties of the substance.

atomic mass unit (amu): the unit used to measure atomic weight. One amu is 1/12 the weight of a carbon-12 atom.

atomic number: the number of protons in the nucleus of an atom.

ATP—see adenosine triphosphate

ATPase (ATP, + *-ase,* enzyme): an enzyme that converts ATP into ADP and in the process releases energy.

ATP synthetase (ATP, + *syn-,* together, + *-the-,* to put, + *-ase,* enzyme): a series of large complex enzymes that catalyze the synthesis of ATP from ADP and inorganic phosphate. The associated reactions are driven by a flow of protons down an electrochemical gradient within mitochondria and chloroplasts. See chemiosmotic theory.

atrium, *pl.* atria (*atri,* entrance hall, + *-um,* structure): (1) a thin-walled chamber in the heart that receives blood from veins. (2) a cavity within the ear of some vertebrates.

auditory (*audit-,* to hear, + *-ory,* place for): pertaining to the sense of hearing, as in the auditory canal.

auricle (pinna) (*aur-,* ear, + *-icle,* little): (1) any ear-like lobe or projection, such as the external ear (pinna) of dogs. (2) a small accessory chamber attached to each atria of the heart.

autecology (*aut-,* self, + *-ec-,* dwelling, + *-logy,* study of): the ecology of individual organisms. Opposite of synecology.

autogamy (*aut-,* self, + *-gam-,* union, marriage, + *-y,* state of): the joining of haploid nuclei within the same organism to restore the diploid condition; may also be referred to as self-fertilization.

autolysis (*aut-,* self, + *-ly-,* to dissolve, *-sis,* process of): self-digestion caused by the action of enzymes produced within the cells being destroyed.

Autonomic Nervous System (ANS) (*aut-,* self, + *-nom-,* name, usage, + *-ic,* pertaining to, + nervous system): a part of the Peripheral Nervous System that controls several involuntary functions, such as respiration, digestion, and heart rate. See Peripheral Nervous System.

autoradiography (*aut-,* self, + *-radi-,* ray, + *-graph-,* to write): a technique used to trace the movement of radioactive substances through an organism; useful in determining the movement and location of photosynthates throughout plants.

autosomes (autosomal chromosomes) (*aut-,* self, + *-som-,* body): any chromosome other than the sex chromosomes (X and Y). Humans have twenty-two pairs of autosomes and one pair of sex chromosomes. Opposite of sex chromosomes.

autotomy (*aut-,* self, + *-tomy,* to cut): self-amputation; as in the self-removal of a damaged appendage by crayfish.

autotroph (*aut-,* self, + *-troph-,* to feed): an organism capable of synthesizing organic material from inorganic materials plus light or chemical energy; such as green plants; also referred to as primary producer. Opposite of heterotroph.

auxiliary cells (*auxil-,* assistance, + *-ary,* pertaining to)—**see subsidiary cells**

auxins (*aux-,* to increase, + *-in,* chemical substance): a group of plant hormones responsible for controlling cell elongation, fruit development, growth movements, and apical dominance.

awn (*awn,* chaff): a brush-like extension of the lemma in grass flowers.
axial, axis (*ax-,* center line, axis, + *-ial,* pertaining to): in reference to the longitudinal midline of the body or other structure; such as the axial skeleton that includes the head, vertebral column, and ribs (if present).
axil (*axill-,* armpit): the upper angle at which the leaf or branch joins the stem.
axillary bud (meristem) (*axill-,* armpit, + *-ary,* place for, + bud): a bud located in the axil of a leaf.
axis—see axial
axon (*axon-,* center line, axis): a long cytoplasmic extension of a neuron that transmits nerve impulses away from the neuron cell body. Opposite of dendrite. See neuron.
axoneme (*axon-,* center line, axis, + *-nem-,* a thread): the central core of microtubules in the cilia or flagella of eukaryotic organisms; arranged in a distinct pattern with nine pairs of microtubules circling one central pair.

B

B lymphocyte (B cell) (*lymph-,* a clear fluid, + *-cyte,* cell): a type of white blood cell that can, when stimulated by an appropriate antigen, become an antibody producing plasma cell. The "B" refers to bone, the place where these cells mature.
bacillus (*bacill-,* a little stick, + *-us,* thing): a rod-shaped bacterium.
background extinction: a natural process of continual, noncatastrophic extinction of species. Opposite of mass extinction.
bacteriophage (*bacter-,* small rod, + *-phage,* to eat): a virus that infects and eventually kills bacteria; also referred to as a phage.
bacterium, *pl.* **bacteria** (*bacter-,* small rod, + *-ium,* region): a unicellular prokaryotic organism; either heterotrophic or autotrophic, and either aerobic or anaerobic.
bar—see megapascal
bark: a collective term for all plant tissue external to the vascular cambium.
Barr body (named after the Canadian anatomist, M. L. Barr): an inactive and condensed X-chromosome appearing as a dense spot in the cell of normal female mammals; most of the genes on this chromosome are not expressed.
basal body (*bas-,* base, + *-al,* pertaining to, + body): a short cylindrical organelle found in animal cells that is associated with anchoring cilia and flagella; similar in structure to centrioles; also referred to as a kinetosome or blepharoplast.
Basal Metabolic Rate (BMR): the minimum number of kilocalories an animal needs at rest; usually determined under controlled conditions.
base: any substance that releases or donates hydroxide ions (OH-) in solution, reducing the hydrogen ion conentration and producing a pH greater than 7. Opposite of acid.
basic research: research that has, at present, no obvious direct application to mankind, but is rather performed for its own sake; such as subatomic particle research.

basidiocarp (*basid-*, pedestal, + *-carp-*, fruit): a spore-bearing structure or "fruiting body" in certain Basidiomycota. Commonly referred to as a "Mushroom."

basidiospore (*basid-*, pedestal, + *-spor-*, spore): a haploid spore produced by certain Basidiomycota, such as the common "mushroom."

basipetalous (*bas-*, base, + *-petal-*, a flower leaf, + *-ous,* pertaining to): a pattern of development in plants starting at the tip and proceeding toward the base of an organ; usually in reference to flowering. Opposite of acropetalous.

basophil (*bas-*, base, + *-phil,* to love): a white blood cell whose cytoplasmic granules readily stain purple-blue with basic dyes such as hematoxylin.

Batesian mimicry (named after the English naturalist, H.W. Bates): a defense mechanism in which a relatively harmless organism evolves the same warning coloration as an organism that is dangerous or poisonous; common in snakes and insects.

beak—see rostrum

behavioral isolation: a prezygotic isolating mechanism that prevents mating between two populations of similar organisms because of differences in mating behavior.

benign (tumor) (*benign,* kindly, + *tum-*, to swell, + *-or,* state of): a cancerous growth that tends not to spread but rather stays in one place. Opposite of malignant.

benthic (*benth-*, depth, + *-ic,* pertaining to): pertaining to the bottom of an aquatic habitat, such as a freshwater lake or ocean.

berry: a fleshy, simple, usually nonsplitting fruit with one or more seeds; such as a banana, a grape, and tomatoe.

beta-carotene—see carotene

beta cells (*beta,* second, + cells): cells in the islets of Langerhans (pancreas) that produce and secrete insulin. See insulin.

beta (β) pleated sheet: one form of secondary structure of proteins formed when two or more polypeptide chains lie side by side and become cross-linked by hydrogen bonds forming a zigzag configuration. See secondary protein structure or proteins.

biennial (*bi-*, two, + *-enni-*, year, + *-al,* pertaining to): a plant that requires two growing seasons to complete its life cycle with vegetative growth occuring in the first year followed by reproductive growth in the second year; such as carrots.

bilateral symmetry (*bi-*, two, + *-lateral-*, side, + *symmet-*, measured together): a body plan in which the right and left sides are approximate mirror images of each other; a distinct "head" end is usually associated with a bilateral organism. Humans are bilateral. The term zygomorphic symmetry is synonomous when describing certain flowers.

bilayer (lipid) (*bi-*, two, + layer): the common two-layered arrangement of phospholipid molecules that make up cellular membranes.

bile (*bil-*, bile): a mixture of organic salts secreted by the liver, stored in the gallbladder, and released into the small intestine; emulsifies fats and aids in fat digestion and absorption.

binary fission (*bin-*, two, + *-ary,* pertaining to, + *fiss-*, to split, + *-ion,* process of): a type of asexual reproduction common to prokaryotic cells in which the parent cell splits into equal halves.

binomial system (of nomenclature) (*bi-*, two, + *-nom-*, name, usage, + *-ial,* pertaining to, + system): the system of assigning organisms a two-part Latin name; the first part is the genus name and the second part is the descriptive (species) name, such as *Homo sapiens.* The genus is always capitalized; the descriptive name is not; both are italicized or underlined. Together the genus and species name is referred to as the scientific name.

bioassay (*bi-*, life, + *-assa-*, trial + *-y*, state of): a quantitative determination of the biochemical properties of a compound; done by testing it under controlled experimental conditions within a living organism.

biodegradable (*bi-*, life, + *-degrad-*, to break down, + *-able,* able to): any substance that can be decomposed by living organisms (bacteria and fungi); such as paper. Opposite of nonbiodegradable.

biogenesis (*bi-*, life, + *-gen-*, origin, + *-sis,* the process of): the fundamental biological principle that states that life only can come from pre-existing life.

biological clock (*bi-*, life, + *-logy,* study of, + *-al,* pertaining to, + clock): a physiological timekeeping mechanism by which organisms adapt to rhythmic external stimuli; such as sleep patterns and day length.

biological control (*bi-*, life, + *-logy,* study of, + *-al,* pertaining to, + control): the use of one organism to specifically control the growth and distribution of another organism; such as using a specific bacteria that only parasitizes and eventually kills specific insects.

biological magnification (*bi-*, life, + *-logy,* study of, + *-al,* pertaining to, + magnification): a process in which certain retained substances become progressively concentrated at each trophic level in a food chain; certain nonbiodegradable herbicides like DDT, or heavy metals, are prime examples.

bioluminescence (*bi-*, life, + *-lumen-*, light, + *-ence,* the condition of): the emission of light produced by living organisms, such as certain Protozoa, and insects, such as the firefly.

biomass (*bi-*, life, + *-mass,* lump, mass): the combined weight of all organisms in a particular habitat.

biomes (*bi-*, life, + *-ome,* mass, group): worldwide groupings of complex communities characterized by unique vegetation and climate; such as the tropical rainforest or arctic biomes.

biosphere (*bi-*, life, + *-spher-*, globe, sphere): the part of the earth that supports life, including land, air, and water; all the earth's biomes.

biotechnology (*bi-*, life, + *-tech-*, techniques, + *-logy,* the study of): the manipulation and application of living organisms, or their components and by-products (DNA and proteins), in commercial or industrial processes. See genetic engineering.

biotic (*bi-*, life, + *-tic,* pertaining to the process of): pertaining to any factor associated with living organisms. Opposite of abiotic.

biotic potential (*bi-*, life, + *-tic,* pertaining to the process of, + *potent-*, to be powerful, + *-ial,* pertaining to): the maximum growth rate a population can experience under ideal (often theoretical) conditions.

biramous (*bi-*, two, + *-ram-*, branch, + *-ous,* pertaining to): with reference to appendages with two distinct branches as opposed to appendages with one or three; such as the appendages in many arthropods. The opposite of uniramous.

bisexual (*bi-*, two, + sex, + *-al,* pertaining to): an organism that can produce both male and female gametes but that usually does not fertilize itself; also referred to as hermaphroditic or monoecious. See monoecious.

bivalent (*bi-*, two, + *-val-*, to be strong, + *-ent,* having the quality of): a pair of homologous chromosomes joined by the centromere during meiosis.

blade (*blade,* leaf): the broad, flat, expanded portion of a plant; usually in reference to leaves but may also refer to algae and lichens.

blastocoel (*blast-*, bud, sprout, + *-coel-*, cavity): the fluid-filled cavity within a blastula. See blastula.

blastocyst (*blast-*, bud, sprout, + *-cyst,* sac): a mammalian blastula. See blastula.

blastomere (*blast-*, bud, sprout, + *-mere,* a part of): an individual cell of the blastula resulting from the division of the zygote.

blastopore (*blast-*, bud, sprout, + *-por-*, opening): the external opening of the archenteron in the gastrula stage of animal development; may become the mouth or anus of the mature organism. See archenteron.

blastula (*blast-*, bud, sprout, *-ula,* little): an embryonic stage in animals that develops immediately after the blastocoel forms; the stage of development that consists of a hollow, fluid-filled sphere of cells; also referred to as blastocyst.

blepharoplast—see basal body

blood: technically considered a type of connective tissue; composed of cells suspended in fluid (plasma) and responsible for carrying nutrients, oxygen, immune system products, and waste materials in the body.

blood-brain barrier: a capillary bed in the brain that controls the passage of most substances into the brain; prevents large fluctuations in brain chemistry.

body cavity—see coelom

bog (*bog,* soft): a distinctive wetland vegetation primarily composed of dead moss, especially of the genus *Sphagnum.*

bordered pit: a cavity in a cell wall in which the secondary wall arches over the pit membrane. Opposite of simple pit. See pit.

boreal (*bore-*, north, + *-al,* pertaining to): pertaining to the floral and fauna of northern regions.

botany (*botan-*, grass, + *-y,* state of): the study of plants.

boundary layer effect: a mechanism whereby transpirational water loss is reduced by establishing a layer of humid air around a leaf. Leaf shape, size, and epidermal hairs (trichomes) aid in maintaining the boundary layer.

Bowman's capsule (named after the British physician, Sir W. Bowman): a double-layered cup of cells at the end of a nephron that surrounds the glomerulus in vertebrate kidneys. The initial process of urine formation occurs here when blood plasma is forced from the capillaries in the glomerulus into Bowman's capsule. See nephron.

bract (*bract-*, thin metal plate): a modified leaf; often greatly reduced, flattened, and associated with flowers.

branchial (*branch-*, gills, + *-ial*, pertaining to): pertaining to the gills or gill region of an animal.

bronchiole (*bronch-*, windpipe, + *-ole*, little): a branch of a bronchus that leads to and forms an alveolus. See alveolus.

bronchus, *pl.* bronchi (*bronch-*, windpipe, + *-us*, thing): one of the first two branches of the trachea that leads into the lung.

brown fat: a type of lipid tissue darkened by a large number of mitochondria; responsible for generating heat in warm-blooded animals; abundant in new borns and used as a fuel source for rapid generation of heat. Opposite of yellow fat.

Brownian movement (named after the Scottish botanist, R. Brown): the motion of small particles and cells in solution caused by water molecules bumping into them and each other.

bryology (*bry-*, moss, + *-logy*, study of): the study of mosses.

buccal (*bucc-*, cheek, mouth, + *-al*, pertaining to): pertaining to the cheek or oral cavity, as in buccal cavity.

bud: (1) an undeveloped shoot located at a terminal or lateral position of a stem. (2) a cytoplasmic protrusion used as a means of asexual reproduction in algae, yeasts, and some bacteria.

budding: a type of asexual reproduction in which protrusions from the parent cell pinch off to form independent but genetically identical cells; common in yeasts and bacteria.

bud scale: a modified leaf surrounding and protecting a bud.

buffer (*buff-*, to blow): a substance that tends to maintain a constant pH of a solution when acids or bases are added.

bulb: a short, underground stem covered with fleshy leaves used as a storage organ; such as onions.

bulbil: a small bulb, or miniature plantlet.

bulliform cells (*bull-*, blister, + *-form*, shape, + cells): large epidermal cells located on the dorsal leaf surface of certain grasses that control the rolling of the leaves during drought.

bundle of His (after the German anatomist, W. His, Jr.): a bundle of nerve and muscle fibers in the vertebrate heart that conducts electrical impulses from the pacemaker (atrioventricular node) to the ventricles.

bundle scar: the mark left on the leaf scar by vascular tissue broken at the time of leaf drop.

bundle sheath: a layer of cells surrounding vascular bundles in leaves that aid in preventing excess water loss from the vascular bundle.

bursa (*burs-*, pouch): a fluid-filled space located in joints that reduces friction between the two bones of the joint.

buttress roots: adventitious roots arising from the stem and used for support; common in monocots, such as corn.

C

C_3 plants (pathway): plants that take up and then, using the Calvin-Benson cycle, incorporate carbon dioxide into a three carbon (C_3) compound (3-phosphoglyceric acid); common to most plants.

C_4 plants (pathway): plants that take up and then incorporate carbon dioxide into a four carbon (C_4) compound (oxaloacetic acid) prior to utilizing the Calvin-Benson cycle; unique to plants growing in hot, dry climates, such as grasses.

caecum—see cecum

calcitonin (*calc-*, calcium, + *-ton-*, condition of, + *-in*, chemical substance): a hormone produced by the thyroid gland that controls the levels of calcium and phosphate ions in the blood.

callus (*call-*, hard skin, + *-us*, thing): (1) undifferentiated tissue formed in response to either mechanical or chemical injury; a common occurance in tissue culture experiments. (2) a horny, thickened layer of skin.

calorie (*calor-*, heat): the amount of heat required to raise the temperature of one gram of water one degree Celsius. The calorie used to measure food energy is the kilocalorie and spelled with a capital **C** (Calorie) and is equal to the amount of heat required to raise the temperature of one kilogram of water one degree Celsius. Both calorie and Calorie are somewhat obsolete terms and have been replaced by the kilojoule (kJ); 1 Calorie is approximately equal to 4.2 kJ. See joule.

Calvin-Benson cycle (Named after M. Calvin and A. Benson, who clarified the process in 1945): the dark reactions of photosynthesis (the light-independent phase) in which carbon dioxide is fixed and converted into glucose.

calyptra (*calytr-*, a veil): a hood-like covering over the developing moss sporophyte capsule.

calyx, *pl.* calices (*caly-*, cup): a collective term for all the sepals of a flower.

CAM—see Crassulacean Acid Metabolism

cambium (*camb-*, exchange, + *-ium*, region): a lateral layer of meristematic cells that divide to provide girth to vascular plants. See vascular cambium and cork cambium.

cancer (*cancer*, a spreading sore, crab): a growth of malignant tissue that has the ability to spread to other regions of the body.

capillary (*capill-*, hairy, + *-ary*, pertaining to): a thin-walled microscopic blood vessel through which cells and molecules are exchanged between the blood and tissues and that connect arterioles to venules.

capillary action: the ''climbing-action'' of water within thin hollow tubes such as xylem cells, in response to the adhesion between the water and xylem surface. See cohesion-adhesion-tension theory.

capitulum (*capit-*, head, + *-ule,* little, + *-um*, structure): a type of inflorescence characteristic of the Compositae family, such as sunflowers. Also known as a head.

capsid (*caps-,* box, + *-id,* tending to): the outer protein coat surrounding the nucleic acid core of a virus.

capsule (*caps-,* box, + *-ule,* little): (1) a gelatinous layer surrounding the cells of blue-green algae and certain bacteria. (2) the sporangium of moss plants. (3) a dry, nonsplitting simple fruit that develops from two or more carpels; such as a cotton capsule.

carapace (*carap-,* shell, + *-ace,* pertaining to): a shell-like plate covering a portion of an animal's body; such as the dorsal portion of a turtle's shell.

carbohydrate (*carbon-,* charcoal, + *-hydr-,* water, + *-ate,* to form): an organic compound containing carbon, hydrogen, and oxygen usually in a 1C:2H:1O ratio; such as sugars, starches, cellulose, and glycogen. See mono-, di-, or polysaccharide.

carbon fixation (carbon, + *fix-*, to put in place, + *-ation*, the process of): the conversion of carbon dioxide into organic compounds like glucose during photosynthesis. This process requires no light and is therefore considered a dark reaction.

carbonyl (carbon, + *-yl*, a chemical radical): a common functional group consisting of a carbon atom double-bonded to an oxygen atom (C=O); common in ketones and aldehydes.

carboxyl (a contraction of carbon and oxygen, + *-yl,* a chemical radical): a common acid radical group characteristic of organic molecules; -COOH.

carcinogen (*carcin-,* cancer, + *-gen,* origin): a substance that causes cancer. See cancer.

cardiac (*cardi-,* heart, + *-ac*, pertaining to): pertaining to the heart, as in cardiac muscle.

carnivore (*carn-,* meat, + *-vor-,* to eat): an animal that eats primarily meat.

carnivorous (*carn-,* meat, + *-vor-,* to eat, + *-ous,* pertaining to): pertaining to animals that eat meat. Opposite of herbivorous.

carotene (*carot-,* carrot): an orange-yellow pigment common in some fruits and flowers.

carotenoid (*carot-,* carrot, + *-oid,* resembling): a group of pigments that includes the xanthophylls (yellow) as well as the carotenes (orange-yellow) that acts as accessory photosynthetic pigments; common in some fruits and flowers.

carpel (*carp-,* fruit): the female reproductive portion of a flower composed of a stigma, style, and one or more ovules.

carrageenin (named after the red algae "carrageen," + *-in,* chemical substance): a polysaccharide extracted from certain red algae; commercially marketed as an emulsifying agent.

carrier-mediated transport—see active transport

carrying capacity: the maximum number of individuals of a specific population that can be supported by the available environmental resources; symbolized K.

cartilage (*cartilag-*, gristle): a flexible type of connective tissue composed of a large number of protein (collagen) fibers; a major component of the skeletal system of vertebrates.

caryopsis (*caryi-*, a nut, + *-opsis*, appearance): a dry, simple, nonsplitting fruit with one seed in which the seed coat is fused to the pericarp; commonly referred to as a grain.

Casparian strip (named after the German botanist, R. Caspary): a waterproof layer of wax forming a band (in a tangential plane) around endodermal root cells: forces water to enter the vascular system rather than remaining in the intercellular spaces.

catabolism (*cata-*, down, + *-bol-*, to put, +, *-ism*, the process of): the sum of all chemical reactions in which large molecules are broken down to form simpler ones and release energy; such as hydrolysis. Opposite of anabolism.

catalase (*cata-*, down, + *-ase*, enzyme): an enzyme that catalyzes the breakdown of hydrogen peroxide to water and oxygen.

catalyst (*cata-*, down, + *-lyst*, to loosen): any substance that regulates the conditions at which a chemical reaction occurs but does not become part of the end product of the reaction; enzymes are biological catalysts.

cation (*cata-*, down, + *-ion*, to wonder): a positively charged ion produced by the loss of one or more electrons; such as Na^+. Opposite of anion.

catkin (*catkin*, little cat): a spike-like group of unisexual flowers found only in some woody plants; a type of inflorescence commonly know as a "pussywillow."

caudal (*caud-*, tail, + *-al*, pertaining to): pertaining to the tail, such as the caudal fin on some fishes.

cauline (*caul-*, stem, + *-ine*, having the character of): arising from or belonging to the stem.

cecum (caecum) (*cec-*, blind end, + *-um*, structure): a blind-ended sac attached to the beginning of the large intestine of some animals; used to store food and enhance digestion and absorption of nutrients.

cell (*cell*, a small room): the basic structural and functional unit of all living organisms; usually microscopic in size. See eukaryotic and prokaryotic.

cell cycle: a series of phases dividing eukaryotic cells pass through including mitosis (M phase), cytokinesis, gap one (G_1 phase), DNA synthesis (S phase), and gap two (G_2 phase). Together G_1, S, and G_2 make up interphase.

cell division—see mitosis, and meiosis

cell membrane (cell, + *membran-*, a coating): a thin molecular bilayer of phospholipids and proteins that surrounds the cytoplasm of all cells and controls the movement of materials into and out of the cell; also referred to as plasma membrane or plasmalemma.

cell plate: the initial stage in the formation of a new cell wall and cell membrane, appearing in late telophase of plant mitosis.

cell theory (cell, + *theor-*, to look at, + *-y*, process): the theory that all living things are made of cells and come from pre-existing cells. This theory has not been disproven since first proposed in 1839 by Schleiden and Schwann.

cell wall: a relatively rigid layer of cellulose and lignin lying outside the cell membrane of all plant cells, some protists, and most prokaryotic cells.

cellular differentiation—see differentiation

cellular respiration (*cellul-*, a small cell, + *-ar*, pertaining to, + *respir-*, to breathe, + *-ation*, the process of): a complex sequence of metabolic reactions including glycolysis, the Krebs cycle, and the electron transport chain, all of which derive energy (ATP) from glucose.

cellulose (*cellul-*, a small cell, + *-ose*, resembling): a highly insoluble structural polysaccharide; the main component of cell walls.

Celsius (named after the Swedish astronomer, A. Celsius): a temperature scale (symbolized °C) that measures the boiling point of water at 100°C and the freezing point of water at 0°C.

Central Nervous System (CNS): usually in reference to the brain and spinal cord of vertebrates but may also apply to similar regions in invertebrates, such as earthworms.

centrioles (*centri-*, center, + *-iol*, little): two cylindrical cytoplasmic organelles unique to animal cells composed of nine microtubular triplets; associated with the formation of the spindle apparatus during mitosis and meiosis. Identical in structure to basal bodies.

centromere (*centri-*, center, + *-mere*, a part of): a constricted region usually at the center of chromosomes at which sister chromatids are held together; spindle fibers are also attached here during mitosis and meiosis.

cephalization (*cephal-*, head, + *-ization*, process of): the evolutionary trend of concentrating structures specialized for feeding and sensing in the "head" or front-end of the animal; common to bilaterally symmetrical animals.

cephalothorax (*cephal-*, head, + *-thorax*, chest): a region of the body common to many Crustacea and Arachnida in which the head and thorax are fused.

cerebellum (*cerebr-*, brain, + *-ellum*, little): the second largest part of the vertebrate brain located beneath the cerebrum; responsible for controlling unconscious muscular activities and balance.

cerebral cortex (*cerebr-*, brain, + *-al*, pertaining to, + *cortex*, bark): the thin outer layer of the cerebrum consisting of a large number of nerve cell bodies; highly developed in mammals and responsible for conscious activities.

cerebrum (*cerebr-*, brain, + *-um*, structure): the largest part of the vertebrate brain consisting of right and left hemispheres; responsible for controlling learning, voluntary muscle action, and most sensory processing.

cervix (*cervix*, neck): the part of any organ resembling a neck; usually in reference to the base of the uterus.

chalaza (*chalaza-*, a small swelling): the base of an ovule or seed where the funiculus fuses with the nucellus and integuments.

chaparral (from the species *chaparra*, "dwarf oak"): a biome characterized by drought-resistant evergreen shrubs and small trees; typical along coastlines exposed to cool ocean currents.

chelicera, ***pl.*** **cheliceratae** (*chel-*, claw, + *-cerat-*, horn): the first pair of appendages in arthropods located anterior to the mouth and used to move food into the mouth; may be modified into fangs or pincers as in spiders.

cheliped (*chel-*, claw, + *-ped-*, foot): the first pair of legs in most crustaceans; often modified into pincers used for capturing and crushing prey.

chemical evolution (*chem-*, chemical, + *-al,* pertaining to, + *evolut-*, an unrolling, + *-ion,* process of): an increase in the molecular complexity of chemical compounds; believed to have preceeded the origin and evolution of life.

chemiosmosis (chemiosmotic theory) (*chem-*, chemical, + *-al,* pertaining to, + *-osmo-*, pushing, + *-sis,* process of): a process associated with cellular respiration and photosynthesis in which the energy released as electrons passed down the electron transport chain is used to create a proton (energy) gradient. The energy released from the flow of protons down the gradient is used to phosphorylate ADP to form ATP; occurs in mitochondria and chloroplasts.

chemoautotrophic (*chem-*, chemical, + *-aut-*, self, + *-troph-*, to feed, + *-ic,* pertaining to): organisms (usually bacteria) that produce their own food from chemical energy and inorganic nutrients. See autotroph. Opposite of heterotroph. Also referred to as chemolithotrophic.

chemoreceptor (*chem-*, chemical, + *-recept-*, to receive): a sensory cell or organ capable of detecting and responding to a chemical stimuli, such as the sensory cells associated with taste and smell.

chemotaxis (*chem-*, chemical, + *-taxis,* ordered movement): a movement toward or away from a chemical stimulus; also referred to as chemotropism.

chemotherapy (*chem-*, chemical, + *-therap-*, care, + *-y,* state of): the use of chemicals in the treatment or control of disease, especially cancer.

chiasma, ***pl.*** **chiasmata** (*chiasm-*, a cross): a region in a tetrad (paired homologous chromosomes) at which crossing over and genetic recombination occur; usually X-shaped. See crossing over.

chimera (chimaera) (a monster from Greek mythology that was part lion, part goat, part snake): (1) a single organism consisting of a mixture of cells from two or more fused zygotes. (2) a hybrid of mixed characteristics produced by grafting. Chimera may occur naturally or artificially.

chitin (*chit-*, tunic, + *-in,* chemical substance): a hard polysaccharide occurring in the cell walls of many fungi as well as the exoskeleton of all arthropods.

chlorenchyma (*chlor-*, green, + *-en-*, in, + *-chym-*, infusion): parenchyma cells containing chloroplasts; typically found in leaves. See parenchyma.

chloroflurocarbons (CFCs): substances once widely used as aerosol propellants and refrigerants. The dissociation of CFCs in the upper atmosphere is known to release chlorine which reacts with and destroys ozone.

chlorophyll (*chlor-*, green, + *-phyll,* leaf): any one of several green photosynthetic pigments necessary for photosynthesis.

chloroplast (*chlor-*, green, + *-plast,* membrane): a membrane-bound organelle containing chlorophyll; specialized for photosynthesis. See plastid.

chlorosis (*chlor-*, green, + *-osis,* diseased): a condition caused by a lack of iron resulting in a loss or reduction in chlorophyll production; a general yellowing.

choanocytes (*choan-*, funnel, + *-cyte,* cell): flagellated cells that line the cavities of sponges; used by sponges as a filtering mechanism to trap food particles; also referred to as collar cells.

cholesterol (*chole-*, bile, + *-ster-*, hard, + *-ol,* denoting an alcohol): a steroid alcohol found in bile, blood, and animal fat and an essential part of animal cell membranes; a precursor to steroid hormones and thought to be associated with certain circulatory problems such as athlerosclerosis.

chorion (*chori-*, skin): the outermost of the four extraembryonic membranes that surrounds the embryos of mammals, birds, and reptiles; in mammals it forms part of the placenta.

chromatid (*chromat-*, color, + *-id,* tending to): one of two identical strands of a chromosome attached at the centromere as a result of DNA replication; chromosome separation occurs at anaphase of mitosis or anaphase II of meiosis. Together, both strands are referred to as sister chromatids.

chromatin (*chromat-*, color, + *-in,* chemical substance): the diffuse thread-like network of DNA, RNA, and nucleoproteins that appears during interphase.

chromatography (*chromat-*, color, + *-graph-*, to write, + *-y,* process): a technique used to separate compounds from a mixture on the basis of their solubility characteristics and molecular size.

chromatophore (*chromat-*, color, + *-phore,* to bear): an animal cell containing pigment and usually located in the skin.

chromoplast (*chrom-*, color, + *-plast,* membrane): a membrane-bound organelle containing pigments other than chlorophyll (carotene); common in fruit and flowers.

chromosomes (*chrom-*, color, + *-som-*, body): rod-shaped bodies of condensed DNA usually visible only during cell division. Chromosomes found in prokaryotic cells are closed loops of DNA, while those in eukaryotic cells are longer, linear strands of DNA complexed with various types of proteins.

chromosome puffs: enlarged, open loops of DNA associated with intense mRNA synthesis; usually seen in the "giant" polytene chromosomes of insects.

chyme (*chym-*, juice, infusion): the semifluid food material that passes from the stomach to the small intestine; produced by the muscular action and digestive enzymes of the stomach.

cilium, ***pl.*** **cilia** (*cili-*, eyelash, + *-um,* structure): a short, hair-like projection on the surface of some eukaryotic cells and used for locomotion and feeding by some unicellular organisms; structurally similar to flagella but much shorter and more abundant. See axoneme.

circadian (*circ-*, around, + *-diem-*, day): a regular biological rhythm of physiological activity repeated approximately every 24 hours, such as leaf or flower movements.

circinate vernation (*circ-*, round, + *-ate,* characterized by having, + *verna-*, sloughing, + *-tion,* the process of): the coiled arrangement of young fern fronds.

cisternae (*cistern-*, box, chest): any spaces or cavities containing fluid; usually in reference to the space between the membranes of endoplasmic reticulum.

cistron: a region of DNA specifying the sequence of a single polypeptide, or functional RNA molecule; a structural gene.

citric acid cycle—see Krebs cycle

cladogenesis (*clad-*, branch, + *-gen-*, origin, + *-sis,* process of): a major pattern of evolutionary change involving one lineage branching into two or more separate lineages with the parent line not becoming extinct; also referred to as splitting evolution.

cleavage (*cleav-*, to divide, + *-age,* collection of): (1) a form of mitosis in animal embryos in which no cell growth occurs between divisions resulting in a multicellular ball of progressively smaller cells. (2) the process of cytokinesis in animal cells, distinguished by pinching of the plasma membrane.

cleistothecium (*cleist-*, close, + *-thec-*, case, + *-ium,* region): a closed ascocarp. See ascocarp.

climax community (*climax,* ladder, + *commun-*, living together, + *-ity,* state of): the final, stable stage in ecological succession dominated by species that tend not to be replaced. See secondary succession.

cline (*clin-*, to lean, sloped): a pattern of gradual variation in characteristics among individuals of a population because of gradual changes in their geographical range.

cloaca (*cloac-*, sewer): a common chamber that receives digestive wastes and urogenital products (urine and gametes); common in most vertebrates except mammals.

clone (*clon-*, twig): a genetically identical organism or population of organisms arising by mitotic cell division from a single ancestor; either naturally or artificially produced.

cnidocyte (*cnid-*, stinging nettle, + *-cyte,* cell): a cell housing a thread-like stinger called a nematocyst; used by cnidarians (''jelly-fish'' and other related organisms) to poison their prey.

coacervates (*coacerv-*, to pile up, + *-ate,* to form): microscopic droplets containing a mixture of complex polymers possessing certain cell-like properties; thought to be possible precursors to true cells. See protocell.

coagulation (*coagul-*, to curdle, + *-ation,* the process of): the formation of a clot, as in blood.

coalescence (*coales-*, to grow together, + *-ence,* the condition of): the union of floral organs of the same whorl, such as sepals to sepals.

cocci (*cocc-*, a berry): a spherical-shaped bacterium.

cochlea (*cochl-*, land snail): a complex coiled structure in the inner ear of mammals, birds, and crocodiles that contains the sense organs for hearing.

codominant (*co-*, together, + *-domin-*, ruling, + *-ant,* that which): a hereditary mechanism in which both alleles for a particular trait are expressed in the heterozyous condition, such as the black and white patterns in Friesian cattle.

codon (*code,* writing tablet, + *-on,* particle): three consecutive nucleotides on mRNA or DNA that code for one amino acid. The fundamental unit of the genetic code.

coelom (*coel-*, cavity): a fluid-filled body cavity completely lined with mesoderm; contains internal organs such as those of the urogenital and digestive systems.

coencytic (*coen-*, shared in common, + *-cyt-*, cell, + *-ic*, pertaining to): a tissue or organism composed of multinucleated cells; such as the hyphae of some fungi, algae, and mosses. Also known as siphonaceous or syncytial.

coenzyme (*co-*, together, + *-en-*, within, + *-zym-*, to ferment): a nonprotein (prosthetic) organic molecule essential for the activation of specific enzymes. Most vitamins as well as NAD^+ and FAD function as coenzymes.

coevolution (*co-*, together, + *-evolut-*, to unfold, + *-ion*, the process of): the simultaneous and often complementary evolutionary change occurring in two or more different species. The evolution of some flowers with their specific insect pollinators is a prime example.

cofactor (*co-*, together, + factor): a nonprotein component essential for the activation of specific enzymes; cofactors are either metallic ions or coenzymes.

cohesion (*cohes-*, to stick together, + *-ion*, the process of): the attraction and binding of similar molecules to each other, such as water molecules bonding to each other to form a column of water within xylem cells. See cohesion-adhesion-tension theory.

cohesion-adhesion-tension theory (*cohes-*, to stick together, + *adhere-*, to stick to, + *tens-*, to stretch, + *-ion*, process of): the proposed mechanism by which water is moved up through the xylem by tension established by transpiration pull, the cohesion of water molecules to one another, and the adhesion of water to xylem cell walls, as opposed to the movement of water due to root pressure alone.

coleoptile (*cole-*, sheath, + *-ptil-*, feather): a protective hollow sheath enclosing the first leaf of grass embryos and seedlings.

coleorhiza (*cole-*, sheath, + *-rhiz-*, root): a protective hollow sheath enclosing the embryonic root of grass embryos and seedlings. Similar in morphology to the coleoptile.

collagen (*coll-*, glue, + *-gen*, origin): a tough, fibrous protein found in abundance in bone and tendon.

collar cell—see choanocyte

collenchyma (*coll-*, glue, + *-en-*, in, + *-chym-*, infusion): a type of tissue composed of elongated support cells with thickened corners; used primarily for mechanical support in plants.

colloid (*coll-*, glue, + *-oid*, resembling): a mixture composed of insoluble particles that remain suspended within a fluid and do not settle out.

columella (*colum-*, column, + *-ella*, little): a central column of sterile cells within the sporangia of some mosses; used to support the sporangium.

commensalism (*com-*, together, + *-mensal-*, table, + *-ism*, the process of): a relationship between two different species that benefits one species without harming the other; a type of symbiosis. See symbiosis.

commissure (*commissur*, connection): a junction formed between two structures or organs, such as a nerve fiber connecting two nerve centers.

community (*commun-*, living together, + *-ity*, state of): all the organisms that live and interact in a particular area or habitat.

companion cells: small specialized cells lying adjacent to seive tube cells within the phloem tissue of flowering plants; associated with the physiological process of moving sucrose into the seive tubes ("sucrose loading").

competition (*com-*, together, + *-pet-*, to seek, + *-tion*, process of): fighting between members of the same or different species for a limited resource such as food, shelter, or territory. The intensity of fighting or interaction is highly variable.

competitive exclusion principle (Gause's principle) (named after the Russian geneticist, G. Gause): no two species can coexist and compete for the same limited resources in the same area. The species that is better adapted in utilizing the available resources will tend to exclude the other.

complement (*comple-* to fill up, + *-ment*, state of being): one of 20 blood proteins that helps various defense mechanisms, such as enhancing phagocytosis or amplifying the inflammatory mechanism.

complete flower: a flower possessing all four floral parts: sepals, petals, stamens, and pistil (carpel). Opposite of incomplete flower.

compound (*compon-*, to put together): a chemical substance made of two or more different kinds of elements; usually in a fixed ratio.

compound eye (*compon-*, to put together, + eye): a complex eye composed of multiple lenses or optical units called ommatidia; common to all arthropods. See ommatidium.

compound leaf (*compon-*, to put together, + leaf): a leaf with a blade divided into several leaflets.

concentration gradient: a difference in the amount of a substance between two regions. For example, the amount of water may often be higher inside a cell than outside resulting in the movement of water out of the cell (down the concentration gradient). See osmosis or diffusion.

conceptacle (*concept-*, to conceive, + *-cle*, little): a chamber in which gametes are produced; common to the brown algae *Fucus*.

conclusion: a statement made following an experiment either accepting or rejecting an hypothesis.

condensation reaction (*condens-*, to press close together, + *-ation*, the process of, + reaction): one common type of anabolic reaction in which two molecules join to form one larger molecule, simultaneously releasing a water molecule. The bonding of amino acids together to form proteins is a condensation reaction.

cone—see strobilus

conidiospore (*coni-*, dust, + *-spor-*, spore): an asexual spore produced at the tip of conidiophores; common to powdery mildews.

conifer (*con-*, cone, + *-fer-*, to carry): any cone-bearing plant; usually used in reference to gymnosperms.

conjugation (*conju-*, to join together, + *-ation*, the process of): a sexual union in certain unicellular organisms in which genetic material is transferred from one cell to another.

connective tissue (*con-*, together, + *-nect-*, to bind, + *-ive*, tending to, + tissue): a type of vertebrate tissue of mesodermal origin composed of cells embedded in a fibrous matrix; used to support and connect organs and other tissues, such as cartilage, bone, and blood.

consumer (*con-*, together, + *-sum-*, to use, + *-er*, one who): a heterotroph that obtains its energy by feeding on other organisms. Herbivores may be considered primary consumers while carnivores may be considered secondary or tertiary (higer-level) consumers.

contractile vacuole—see vacuole

controlled experiment: an experiment in which test (experimental) data are compared to control (standard) data to determine the validity of the experiment.

convergent evolution (*converg-*, to come together, + *-ent*, that which, + *evolut-*, an unrolling, + *-ion*, process of): the independent development of similar traits between two or more groups of unrelated organisms as a result of occupying similar ecological niches; such as bats and birds.

copulation (*copul-*, to bond, + *-ation*, the process of): sexual union (intercourse) during which sperm cells are transferred from the male to the female; also referred to as coitus.

corium—see dermis

cork: a suberized layer of dead cells developed from the cork cambium that acts as a barrier to the passage of water into and out of the plant. Also called phellem. See suberin.

cork cambium (cork, + *camb-*, exchange, + *-ium*, region): a lateral meristem that produces phelloderm toward the inside of the plant and cork toward the outside. Also called phellogen and is typical of woody plants.

corm (*corm-*, trunk of a tree): a fleshy, short, vertical, underground stem lacking leaves and used as a storage organ; such as gladiolus "bulbs."

cornea (*corn-*, horny, keratinized tissue): the outer, transparent covering of the eyeball.

corneum (*corn-*, horny, keratinized tissue, + *-um*, structure): a layer of dead keratinized cells; the outermost layer of the epidermis; also referred to as the stratum corneum.

corolla (*coroll-*, little crown): a collective term for all the petals of a flower.

corpus callosum (*corpus-*, body, + *call-*, hard skin, + *-um*, structure): a tight bundle of myelinated nerve fibers connecting the right and left cerebral hemispheres.

corpus luteum (*corpus-*, body, + *lute-*, yellowish, + *-um*, structure): a mass of endocrine cells formed within an ovarian follicle following ovulation and responsible for secreting progesterone and estrogen, which maintain the uterine lining during pregnancy.

cortex (*cortex*, bark): (1) the outer layer of some organs, such as the kidney's renal cortex or the brain's neocortex. (2) a region of cells in stems and roots between the vascular tissue and the epidermis; often a site of food or water storage in plants.

corymb (*corymb-*, a cluster of flowers): an inflorescence with a flat top; characteristic of the carrot family.

cotransport (*co-*, together, + *trans-*, across, + *-port,* to carry): the active transport of one substance against (up) a concentration gradient by simultaneously attaching to another substance moving with (down) a concentration gradient.

cotyledon (*cotyle-*, cup): the first leaf (leaves) produced by the embryo of seed plants. Also known as a seed leaf.

covalent bond (*co-*, together, + *-val-*, to be strong, + *-ent,* that which, + bond): a relatively strong chemical bond formed by the sharing of electrons between atoms.

Crassulacean Acid Metabolism (CAM): a form of carbon metabolism associated with photosynthesis in which the stomata open only at night to allow carbon dioxide uptake, thereby preventing excess water loss by daytime transpiration. During the day the stored carbon is released and photosynthesis is completed. Common in some desert succulants and some grasses. Named after the family Crassulaceae that makes extensive use of this metabolism.

crista, ***pl.*** **cristae** (*crist-,* a ridge, crest): a short, finger-like projection formed by the folding of the inner membrane of mitochondria; the location of the electron transport chain and where ATP is synthesized.

crossing over: the random exchange of corresponding pieces of genetic material between chromatids of homologous chromosomes during synapsis of prophase I, resulting in gene recombination and providing an essential source of genetic variation.

cross section: a section made perpendicular to the longitudinal axis. Also called transverse section. A cross section of a typical plant stem will appear round.

crustose (*crust-,* encrusted, + *-ose,* resembling): a growth form (habit) resembling an encrustation; common to some types of lichens.

cryptic coloration (*crypt-*, hidden, + *-ic,* pertaining to, + coloration): a type of camouflage that allows potential prey to blend in with their surroundings and be difficult to see.

cryptogam (*crypt-*, hidden, + *-gam,* marriage): a seedless vascular plant, such as a fern.

ctenoid scales (*cten-*, comb, + *-oid,* resembling): thin overlapping scales possessing tooth-like spines on the exposed posterior surfaces; characteristic of more advanced fishes.

cultivar (a contraction of cultivated and variety): a strain or breed developed artificially by human manipulation. Virtually all agriculturally important crops are cultivars.

cuticle (*cut-*, skin, + *-cle,* little): (1) an external, cutinized layer covering the outer wall of plant epidermal cells that aids in reducing water loss through transpiration. (2) another name for the epidermis of animals. See cutin.

cutin (*cut-*, skin, + *-in,* chemical substance): a waxy lipid used as a waterproofing material on plant epidermal cells; the major component of the cuticle.

cyanophycean starch (*cyan-*, blue, + *-phyc-*, algae, + *-an*, pertaining to + starch): a polysaccharide used for food storage in blue-green algae (Cyanobacteria); similar to amylopectin.
cyclic AMP (cAMP): a type of adenosine monophosphate with a ring-shaped phosphate group that acts as a second (intermediate) messenger for several vertebrate hormones and neurotransmitters; also functions in chemical communications in slime molds, and acts as a regulator of the *lac* operon.
cyclic photophosphorylation (*cycl-*, circle, +*-ic,* pertaining to + *photo-*, light, + phosphorous, + *-yl*, chemical radical, + *-ation*, the process of): the synthesis of ATP during a cyclic flow of electrons within the thylakoid membranes of chloroplasts; the process is driven solely by light energy.
cycloid scales (*cycl-*, circle, + *-oid*, resembling): thin overlapping scales that have smooth rounded posterior surfaces; characteristic of more primitive fishes.
cyclosis (*cycl-*, circle, + *-sis*, the process of): the circular flow of cytoplasm within a cell to equally distribute nutrients and other materials including organelles; also referred to as cytoplasmic streaming.
cyme (*cyme*, a young sprout): a branching, flat-topped inflorescence in which the apical flower on each branch blooms first.
cyst (*cyst-*, bag): a closed sac, vesicle, or bladder often protected by a thick outer covering.
cytochrome (*cyt-*, cell, + *-chrom-*, color): a large class of iron (heme) containing proteins that function as electron carriers in cellular respiration and photosynthesis.
cytokinesis (*cyt-*, cell, + *-kine-*, to move, + *-sis*, the process of): division of the cytoplasm into two daughter cells following nuclear division (karyokinesis).
cytokinins (*cyt-*, cell, + *-kine-*, movement, + *-in*, chemical substance): a group of plant hormones that helps control cell division, bud growth, and senescence.
cytology (*cyt-*, cell, + *-logy*, the study of): the study of cells.
cytoplasm (*cyt-*, cell, + *-plasm*, formed): the living material within a cell, excluding the nucleus.
cytoplasmic streaming—see cyclosis
cytosine (*cyt-*, cell, + ine, having the character of): one of four nitrogenous nucleotide bases that is a component of nucleic acids (DNA and RNA); complimentary to guanine.
cytoskeleton (*cyt-*, cell, + *-skelet-*, a dried hard body): a three-dimensional framework of microtubules and microfilaments throughout the cytoplasm of cells; provides support and shape to cells and support for organelles.
cytosol (*cyt-*, cell, + *-sol* solution): the semifluid, viscous portion of the cytoplasm.

D

Dalton (named after the English chemist, J. Dalton): a somewhat obsolete unit used to measure the atomic mass of atoms and subatomic particles. The kilogram has largely replaced the Dalton. One Dalton is approximately equal to 1.0×10^{-21} k.
dark reaction—see carbon fixation

Darwinism (named after the English naturalist, C. Darwin): the theory of evolution by means of natural selection. See natural selection.

day-neutral plants: plants whose flowering is not affected by photoperiodism. See long-day plants and short-day plants.

deamination (*de-*, to remove, + *-amine-*, resinous gum, + *-ation*, the process of): the removal of an amino group, usually from an amino acid.

decarboxylation (*de-*, to remove, + carbon, + oxygen, + *-yl-*, a chemical radical, + *-ation*, the process of): the removal of a carboxyl group, usually in the form of carbon dioxide, from an organic compound.

deciduous (*decidu-*, falling off, + *-ous*, pertaining to): the shedding of all leaves at one time; usually at the end of a growing season.

decomposers (*de-*, from, + *-compon-*, to put together): organisms that absorb nutrients from living or nonliving organic material; such as bacteria and fungi.

dehiscent (*dehisc-*, to split, + *-ent*, performing the action of): usually in reference to fruit or other organs that split open at maturity; such as the fruit of a legume (pod). Opposite of indehiscent.

dehydration synthesis (*de-*, to remove, + *-hydr-*, water, + *-ation*, the process of, + *syn-*, to put together, + *-the-*, to place, + *-sis*, process of): a common type of anabolic reaction in which two molecules join to form one larger molecule by removing components that form a water molecule. The formation of a disaccharide from two monosaccharides is a dehydration synthesis. Also referred to as condensation reaction. Opposite of hydrolysis.

deme (*dem-*, people): an interbreeding group of individuals within a specific species.

denature (*de-*, to remove, + nature): to destroy the physical integrity and 3–D configuration of a compound (usually in reference to proteins or nucleic acids) by heat, extreme pH, or chemical treatment, resulting in a loss of metabolic function.

dendrite (*dendr-*, tree, + *-ite*, part of): a short, multibranched extension of a neuron that transmits nerve impulses toward the neuron cell body. Opposite of axon. See neuron.

deoxyribonucleic acid (DNA) (*de-*, to remove, + *-oxy-*, oxygen, + ribose, a pentose sugar, + *-nucle-*, nucleus, kernel, + *-ic*, pertaining to, + acid): a nucleic acid that functions as the genetic material of all organisms; composed of two complimentary strands of nucleotides coiled together into a double helix.

depolarization (*de-*, to remove, + *-pol-*, pole, + *-ization*, to make): the reversal of a cell membrane's electrical properties toward a nonpolarized condition. See sodium-potassium pump.

dermal tissue (system) (*derm-*, skin, + *-al*, pertaining to, + tissue): one of three fundamental tissue systems in plants that gives rise to the epidermis. See ground tissue and vascular tissue.

dermis (*derm-*, skin): the layer of skin below the epidermis; also referred to as corium.

desmosome (*desm-*, to bond, + *-som-*, body): a type of intercellular junction that acts to "weld together" the membranes of adjacent cells; often referred to as a button-like plaque.

desmotubule (*desm-*, to bond, + *-tub-*, tube, + *-ule*, little): a continuation of endoplasmic reticulum between two adjacent plant cells through a connecting plasmodesmata.

determinate cleavage (*determin-*, to end completely, + *-ate*, characterized by having, + *cleav-*, to divide, + *-age*, collection of): a type of embryonic development in animals in which the fate of each embryonic cell is determined very early in development; usually a spiral pattern of cleavage characteristic of protostomes. See spiral cleavage.

determinate growth (*determin-*, to end completely, + *-ate*, characterized by having, + growth): (1) a growth pattern characteristic of animals that stops after the animal has reached a specific size. (2) growth of limited duration due to the loss of meristematic activity; such as flower and leaf growth. Opposite of indeterminate growth.

detritivore (*detrit-*, worn off, + *-vor-*, to eat): a heterotroph that obtains its energy by feeding on dead or decaying organisms as well as waste materials; such as an earthworm, some insects, and bacteria.

detritus (*detrit-*, worn off, + *-us*, thing): dead organic material.

deuterostome (*deuter-*, second, + *-stom-*, mouth): an animal in which the embryo's anus develops from or near the blastopore and the mouth develops from a secondary opening, and in which cleavage is indeterminate. One of two evolutionary lines of coelomate animals that includes echinoderms and chordates. Opposite of protostome.

diakinesis (*dia-*, across, + *-kine-*, motion, + *-sis*, process of): the fifth stage of prophase 1 of meiosis in which the final condensation of the chromosomes occurs.

diaphragm (*dia-*, across, + *-phragm*, fence): a sheet-like muscle separating the thoracic cavity from the abdominal cavity; functions in filling and emptying the lungs through its contractions.

diastole (*diastol-*, to expand): a period when the heart muscle relaxes and the chambers fill with blood. Opposite of systole.

dichotomous (*dich-*, two, + *-tom-*, to cut, + *-ous*, pertaining to): the branching or splitting of an organ or structure into two equal halves.

dicotyledon (*di-*, two, + *-cotyle-*, cup): a group of flowering plants (Dicotyledonae) with embryos having two cotyledons. Often abbreviated dicot.

dictyosome (*dicty-*, a net, + *-som-*, body): another name for the net-like Golgi complex in plant cells. See Golgi complex (apparatus).

dictyostele (*dicty-*, a net, + *-stel-*, a pillar): separate bundles of xylem and phloem forming an incomplete ring around a central core of pith. See siphonostele.

diencephalon (*dien-*, between, + *-cephal-*, brain): the posterior portion of the vertebrate forebrain containing the thalamus, hypothalamus, and pituitary gland; lying between the mesencephalon (midbrain) and telencephalon.

differential permeability—see semipermeable

differentiation (*di-*, apart, + *-fer-*, to carry, + *-ation*, the process of): the irreversible developmental process that leads toward more mature (specialized) cells or tissues.

diffusion (*diffus-*, to pour out, + *-ion*, the process of): the passive and spontaneous movement of molecules from a region of high concentration to a region of low concentration (down a concentration gradient); a movement of molecules driven by the kinetic energy of the molecules involved.

digestion (*digest-*, separating out, + *-ion*, process of): the process of breaking down complex foods into molecules small enough to be absorbed by the body.

dihybrid (*di-*, two, + *-hybrid-*, mixed offspring): the offspring of parents who differ in two distinct genetic traits; genotypically heterozygous (Aa / Bb).

dihybrid cross (*di-*, two, + *-hybrid-*, mixed offspring, + cross): a genetic cross between two parents who differ in two distinct traits.

dikaryotic (*di-*, two, + *-kary-*, nut, + *-tic*, pertaining to the process of): cells or organisms with two unfused nuclei; a condition common to certain fungi.

dimer (*di-*, two, + *-mer*, a part of): a molecule composed of two similar parts.

dimorphic (*di-*, two, + *-morph-*, shape, *-ic*, pertaining to): organisms possessing two distinct forms, such as ferns with fertile and sterile fronds.

dioecious (*di-*, two, + *-eci-*, dwelling, + *-ous*, pertaining to): having male and female sex organs in different individuals of the same species. Also known as unisexual when referring to animals. Opposite of monoecious.

diploblastic (*dipl-*, double, + *-blast-*, sprout, bud, + *-ic*, pertaining to): pertaining to animals with two embryonic germ layers, endoderm and ectoderm; such as flatworms (Platyhelminthes). See triploblastic.

diplococcus (*dipl-*, double, + *-cocc-*, a berry, + *-us*, thing): an arrangement of cells in clusters of two; usually pertaining to spherical (coccus) bacteria.

diploid (*di-*, two, + *-ploid*, mulitple of): a normal condition in which a cell or organism has two sets of chromosomes (one from each parent); symbolized 2n. Opposite of haploid.

diplotene (*dipl-*, double, + *-ten*, to hold): the fourth stage of prophase I in meiosis during which chiasmata appear between nonsister chromatids.

directional selection: a form of natural selection that favors those individual organisms at the extreme ends of the phenotypic range. Also referred to as diversifying selection.

disaccharide (*di-*, two, + *-sacchar-*, sugar, + *-ide*, denoting a chemical compound): a type of carbohydrate (a sugar) made of two monosaccharides; such as sucrose, which is made of glucose and fructose.

disk flower (floret): a small tubular flower that is radially arranged in the center of most Compositae inflorescences; such as a sunflower.

dissociate (*dis-*, apart, + *-soci-*, to join together, + *-ate*, characterized by having): usually in reference to compounds breaking down into simpler molecules, or electrolytes dissolving in solution, yielding ions.

distal (*dist-*, apart, remote, + *-al*, pertaining to): pertaining to a structure located away from its point of attachment or origin. Opposite of proximal.

divergent evolution—see adaptive radiation

diversifying selection—see directional selection

DNA—see deoxyribonucleic acid

dominant (allele) (*domin-*, ruling, + *-ant*, that which, + *allel-*, of one another): an allele (gene) that is always expressed in the phenotype, regardless of whether it is homozygous or heterozygous and that will mask the action of the recessive allele of the same gene. See recessive allele.

dormancy (*dorm-*, to sleep, + *-ancy*, state of being): a period of inactivity characterized by a general reduction of metabolic activities; usually induced by a specific combination of environmental factors including temperature, light, and moisture; such as seed or bud dormancy.

dorsal (*dors-*, back, + *-al*, pertaining to): pertaining to the back or posterior part of an organ or body. Opposite of ventral.

dorsiventral symmetry (*dors-*, back, + *-ventr-*, belly, + *-al*, pertaining to, + *symmet-*, measured together): pertaining to an organism that has distinct dorsal and ventral surfaces; such as pea flowers and some flatworms. See bilateral symmetry.

double fertilization: a process unique to flowering plants in which one of two sperm nuclei fuses with an egg nucleus to form the zygote, and the other sperm nuclei simultaneously fuses with two polar nuclei to form triploid endosperm tissue.

double helix (double, + *helix*, spiral, coil): the term used to describe the "spiral-staircase" shape of the DNA molecule; in reference to the two polynucleotide molecules wound into a 3–D spiral-like shape.

Down's syndrome (after the English physician, J. Langdon-Down): a genetic disorder in humans associated with an extra chromosome number 21; characterized by mental retardation and developmental abnormalities; also referred to as trisomy 21.

drupe (*dryppa*, olive): a fleshy, simple, nonsplitting, oneseeded fruit, such as an olive or peach.

druse (*dru-*, bump): a spherical calcium oxalate crystal with many projections. Abundant in pear fruit.

dulse (*dulc-*, sweet): a commercially marketed preparation of dried red algae and usually sweet depending on the species.

duodenum (*duoden-*, twelve, + *-um*, structure): the first (upper) portion of the small intestine in vertebrates into which ducts from the pancreas and gallbladder enter; so named because of its length, approximately 12 fingers' breadth. Primary function is the digestion and absorption of food.

dynein (*dyn-*, power, + *-in*, chemical substance): a contractile protein forming portions of the microtubules of cilia and flagella.

E

ecdysis (*ecdy-*, slipping out, + *-sis*, process of): shedding the dead outer layer of skin (cuticle); a process common to insects and crustaceans; also referred to as molting.

ecological isolation (*ec-*, dwelling, + *-logy,* study of, + *-al,* pertaining to, + isolation): a situation in which gamete exchange between two groups of organisms within the same geographical area is prevented because of genetic differences that have arisen due to a particular feature of the environment.

ecological niche (*ec-*, dwelling, + *-logy,* study of, + *-al,* pertaining to, + *nich,* to nest): a description of an organism's "occupation" within its community; how an organism interacts with its environment.

ecological pyramid (*ec-*, dwelling, + *-logy,* study of, + *-al,* pertaining to, + pyramid): a diagram representing the quantitative relationships of biomass, numbers of organisms, or energy levels at each trophic level in a ecosystem, beginning with primary producers at the base. Depending on the variable being measured, the pyramid may be referred to as a pyramid of biomass, pyramid of numbers, etc.

ecological succession (*ec-*, dwelling, + *-logy,* study of, + *-al,* pertaining to, + *success-,* to follow, + *-ion,* process of): the gradual change of species composition within a community from simpler to more complex. See primary succession and secondary succession.

ecology (*ec-*, dwelling, + *-logy,* study of): the study of how organisms interact with their environment.

ecosystem (*ec-*, dwelling, + *-system,* a composite whole): a community of organisms along with all associated biotic and abiotic factors that maintain a stable system. See community.

ectoderm (*ecto-,* outer, + *-derm,* skin): the outermost of the three embryonic germ layers of animals that gives rise to the nervous system and the integument.

ectoparasite (*ecto-,* outer, + *-para-,* along side, + *-sit-,* food): an organism that lives on the external surface of its host; such as a tick or louse. Opposite of endoparasite.

ectophloic (*ecto-,* outer, + *-phlo-,* bark, + *-ic,* pertaining to): an arrangement of vascular tissue with phloem external to xylem.

ectoplasm (*ecto-,* outer, + *-plasm,* formed): the clear, nongranular portion of the cytoplasm just inside the cell membrane. Opposite of endoplasm.

ectotherm (*ecto-,* outer, + *-therm,* heat): an animal that obtains most of its body heat from the external environment; such as a reptile, amphibian, and a fish; also referred to as cold-blooded or poikilothermic. Opposite of endotherm.

edema (*edem-,* to swell): an abnormal swelling of tissue due to the accumulation of fluid in the intercellular spaces.

effector (*ef-,* out of, away, + *-fect-,* to make, + *-or,* result of the act of): a gland that is stimulated to secrete or a muscle that is stimulated to contract in direct response to a nerve impulse.

efferent (*ef-,* out of, away, + *-fer-,* to carry, + *-ent,* performing the action of): pertaining to a structure carrying something away from or out of another organ or region; such as blood vessels and nerves. Opposite of afferent.

egg: a nonflagellate female gamete; also referred to as ovum. See ovum.

elaters (*elate-*, driver): dead, elongated cells that coil and uncoil in response to changes in humidity and aid in the dispersal of spores; common in mosses and primitive vascular plants.

electrolyte (*electr-*, amber, + *-lyt-*, dissolve): a substance that dissociates into ions when dissolved in a solvent (usually water) and is able to conduct an electrical current.

electron (*electr-*, amber, *-on*, a particle): a primary subatomic particle with a negative charge; located in orbitals (shells) around the nucleus of an atom.

electron acceptor: a molecule that will accept one or more electrons in an oxidation-reduction reaction, becoming reduced in the process. In the Krebs cycle, NAD^+ and FAD accept electrons forming NADH and $FADH_2$ respectively. Opposite of electron donor.

electron carrier: a molecule that can alternately gain or lose electrons becoming reduced or oxidized; such as a cytochrome molecule.

electron donor: a molecule that will give up one or more electrons in an oxidation-reduction reaction, becoming oxidized in the process. NADH and $FADH_2$ donate electrons forming NAD^+ and FAD respectively. Opposite of electron acceptor.

electron transport chain (system): the third and final stage of cellular respiration in which ATP is synthesized by means of an exergonic movement of electrons. This system occurs in the inner membrane of mitochondria; also referred to as oxidative phosphorylation. See chemiosmosis.

electrophoresis (*electr-*, amber, + *-phore-*, to bear, + *-sis*, process of): a technique used to separate compounds from a mixture on the basis of their size and electric charge by measuring their movements through an electrified medium (gel).

element (*element-*, first principle): a substance that cannot be broken down by ordinary chemical means; the simplest of substances made up of only one type of atom.

embryo (*em-*, in, + *-bry-*, to swell, grow): an early developmental stage of an organism that takes place within an egg, seed, or body of its mother. In human development, the embryo stage occurs from the time of conception to the eighth week of pregnancy.

embryo sac (*em-*, in, + *-bry-*, to swell, grow, + sac): the mature female gametophtye of angiosperms usually containing eight haploid nuclei.

emigration (*e-*, out, + *-migr-*, to move, + *-ation*, the process of): the process of leaving one location and moving to another to live. Opposite of immigration.

endemic (*en-*, in, + *-dem-*, people, country, *-ic*, pertaining to): pertaining to an organism that is unique or native to a specific region or country; not introduced.

endergonic reaction (*endo-*, within, + *-erg-*, work + *-ic*, pertaining to, + reaction): a chemical reaction that requires an input of energy; a nonspontaneous ("uphill") process. Opposite of exergonic reaction.

endocarp (*endo-*, within, + *-carp*, fruit): the innermost of the three layers of the pericarp (fruit wall) of angiosperms.

endocrine gland (*endo-*, within, + *-crine,* to secrete): a ductless gland that secretes its product directly into tissue or blood; such as the thyroid gland. Opposite of exocrine gland.

endocytosis (*endo-*, within, + *-cyt-*, cell, + *-sis,* the process of): a form of active transport in which localized regions of the cell membrane invaginate to form a vacuole engulfing the material to be taken into the cell. The vacuole eventually pinches off to become a cytoplasmic vesicle. See phagocytosis and pinocytosis.

endoderm (*endo-*, within, + *-derm,* skin): the innermost of the three embryonic germ layers of animals that gives rise to the lining of the gut as well as most visceral organs.

endodermis (*endo-*, within, + *-derm-*, skin): the innermost layer of the cortex, a portion of which is suberized by the Casparian strip; common in the roots of advanced land plants. See Casparian strip.

endogenous (*endo-*, within, + *-gen-*, origin, + *-ous,* pertaining to): produced or caused by internal factors, such as naturally produced hormones. Opposite of exogenous.

endomembrane system (*endo-*, within, + *-membran-*, a coating, + system): a collection of membranes within eukaryotic cells that are related either by direct contact with each other or through the transfer of membrane segments and fusion of vesicles; includes the nuclear membrane, endoplasmic reticulum, Golgi apparatus, various vacuoles, and the cell membrane.

endometrium (*endo-*, within, + *-metr-*, mother, uterus, + *-ium,* region): the lining of the uterus that is richly supplied with blood vessels; the site of embryo implantation. Opposite of myometrium.

endoparasite (*endo-*, within, + *-para-*, along side, + *-sit-*, food): an organism that lives in or on the internal organs of its host; such as an intestinal worm. Opposite of ectoparasite.

endophyte (*endo-*, within, + *-phyt-*, plant): a plant or plant-like organism growing within another plant; such as certain fungi. Opposite of epiphyte

endoplasm (*endo-*, within, + *-plasm-*, formed): the granular portion of the cytoplasm that surrounds the nucleus. Opposite of ectoplasm.

endoplasmic reticulum (ER) (*endo-*, within, + *-plasm-*, formed, + *-ic,* pertaining to, + *reticul-*, network, + *-um,* structure): an extensive network of membranes, tubules, and vesicles within eukaryotic cells; may be studded with ribosomes (rough ER) and aid in protein synthesis or be ribosome free (smooth ER) and synthesize lipids.

endorphins (a contraction of endogenous and morphine): a group of naturally synthesized opiate-like (morphine) hormones produced in the brain that act as natural painkillers.

endoskeleton (*endo-*, within, + *-skelet-*, a dried hard body): an internal framework used for support and the attachment of muscles; characteristic of vertebrates and echinoderms; usually composed of cartilage and bone. Opposite of exoskeleton.

endosperm (*endo-*, within, + *-sperm*, seed): a highly nutritive food reserve resulting from the process of double fertilization; found only in certain angiosperm seeds. See double fertilization.

endospore (*endo-*, within, + *-spor-*, spore): a thick-walled dormant spore formed within the parent cell of bacteria and some blue-green algae.

endosymbiont theory (*endo-*, within, + *-sym-*, together with, + *-bi-*, life, + *ont*, individual, + theory): a current idea explaining the origin and evolution of eukaryotic cells. This theory maintains that certain organelles such as mitochondria and chloroplasts originated as symbiotic prokaryotes living within larger prokaryotes.

endothelium (*endo-*, within, + *-thel*, nipple, + *-ium*, region): the innermost layer of cells lining blood vessels; composed of simple squamous epithelium.

endotherm (*endo-*, within, + *-therm*, heat): an animal (all birds and mammals) that produces and maintains its own body temperature through various metabolic processes; also referred to as warm-blooded or homeothermic. Opposite of ectotherm.

endotoxin (*endo-*, within, + *-tox-*, a poison, + *-in*, chemical substance): a toxin produced within certain bacteria and released after the cell dies and ruptures; responsible for such symptoms as fever and muscle aches.

enterocoel (*enter-*, gut, + *-coel-*, cavity): a coelom that develops from mesodermal outpockets that eventually separate to form sacs and eventually a coelom; characteristic of deuterostomes. See deuterostome.

enteron (*enter-*, gut): the digestive cavity.

entropy (*entrop-*, transformation, + *-y*, state of): a quantitative measure of disorder or random order (energy that is not available to an organism); symbolized S.

enzyme (*en-*, in, + *-zym-*, ferment): a protein catalyst that controls the speed and direction of chemical reactions within living organisms without being used up by the reaction.

eosinophil—see acidophil

epaxial (*epi-*, upon, + *-ax-*, axis, + *-al*, pertaining to): usually pertaining to muscles located above or dorsal to the vertebral column. Opposite of hypaxial.

ephemeral (*ephem-*, short-lived, + *-al*, pertaining to): organisms that complete their life cycle in a short period of time, usually several days to a few weeks; such as certain alpine plants, or certain insects.

epicotyl (*epi-*, upon, + *-cotyle-*, cup): the portion of a seedling above the cotyledons and below the first true leaf.

epidermis (*epi-*, upon, + *-derm-*, skin,): the outermost cell layer of an organism that acts as a protective layer; also referred to as cuticle.

epididymis, *pl.* epididymides (*epi-*, upon, + *-didym-*, testicle, + *-us*, thing): a long convoluted tube located on the posterior surface of a testicle in which sperm cells mature before passing into the vas deferens.

epigenesis (*epi-*, upon, + *-gen-*, origin, + *-sis*, process of): the development of form in embryos.

epinephrine (*epi-*, upon, + *-nephr-*, kidney, + *-in*, chemical substance): a hormone produced by the inner portion (medulla) of adrenal glands that regulates blood sugar levels, blood pressure, and heartbeat rate; stimulates the "fight or flight" response; also referred to as adrenaline.

epiphyte (*epi-*, upon, + *-phyt-*, plant): a nonparasitic plant living on another plant for support; commonly referred to as an air plant; such as an orchid. Opposite of endophyte.

epistasis (*epi-*, upon, + *-sta-*, to control, + *-sis*, process of): the process in which one gene alters or inhibits the expression of another gene at another locus.

epitheca (*epi-*, upon, + *-thec-*, case): the outer, larger half of the siliceous portion of a diatom cell wall. Opposite of the hypotheca.

epithelium, *pl.* epithelia (*epi-*, upon, + *-thel-*, nipple, + *-ium*, region): a type of vertebrate tissue primarily arranged in layers; covers external body surfaces, lines cavities, and forms glands; also referred to as epithelial tissue.

equilibrium (*equ-*, equal, + *-libr-*, balance, +, *-ium*, part)—**see homeostasis**

erythrocyte (*erythr-*, red, + *-cyte*, cell): a red blood cell that functions in the transport of oxygen and carbon dioxide.

essential compounds: any substance an organism needs to stay alive but is unable to synthesize. In adult humans, the fatty acids linolenic and linoleic and eight amino acids are essential.

ester bond: a chemical bond between a carboxylic acid and an alcohol.

estivation (*estiv-*, summer, + *-ation*, the process of): summer dormancy brought on by heat and lack of water.

estrogens (*estr-*, mad desire, + *-gen-*, origin): a collective term for female sex hormones that maintain secondary female sex characteristics. Opposite of androgens.

estrus (*estr-*, mad desire, + *-us*, thing): the cyclic period of sexual receptivity occurring in female mammals having an estrous cycle; usually occurs during ovulation of the egg; also referred to as "in heat."

ethology (*eth-*, character, + *-logy*, study of): the study of how animals behave in their natural environment, with an emphasis on the adaptive value of behavior.

ethylene: a gaseous hydrocarbon ($H_2C = CH_2$) that acts as a plant hormone promoting fruit ripening and leaf drop.

etiolation (*etiol-*, pale, + *-ation*, the process of): a condition characteristic of plants grown in total darkness involving a lack of chlorophyll, poor leaf development, and increased stem length.

etiology (*eti-*, cause, + *-logy*, study of): the study of the cause of a disease.

euchromatin (*eu-*, true, + *-chromat-*, color, + *-in*, chemical substance): loosely coiled chromatin that is actively being transcribed; stains less intense than heterochromatin. See heterochromatin.

eucoelomate (*eu-*, true, + *-coel-*, cavity, + *-ate*, to form): an animal possessing a true body cavity (coelom) between the gut and body wall; such as a vertebrate. Opposite of acoelomate. See coelom.

eukaryotic (eucaryotic) (*eu-*, true, + *-kary-*, nut, nucleus, + *-tic*, pertaining to the process of): an organism possessing membrane-bound organelles such as a nucleus. All plants, fungi, and animals are eukaryotic. Opposite of prokaryotic.

euploid (*eu-*, true, + *-ploid*, multiple of): the loss or gain of one or more entire sets of chromosomes; the most common type is polyploid. See polyploid.

eustele (*eu-*, true, + *-stel-*, pillar): an arrangement of vascular tissue in which there are distinct bundles of xylem and phloem around a pith; common in gymnosperms and angiosperms.

eutrophication (*eu-*, true, + *-troph-*, to feed, + *-ation*, process of): a process usually occurring in lakes with abnormally high rates of biological productivity supported by elevated nutrient levels; caused by pollution or natural processes.

evolution (*evolut-*, an unrolling, + *-ion*, process of): a change in the gene frequency within a population from generation to generation. See microevolution and marcoevolution.

excretion (*ex-*, external to, + *-cret-*, to separate, + *-ion*, process of): the elimination of nitrogenous wastes, such as urine.

exergonic reaction (*ex-*, external to, + *-erg-*, work, + *-ic*, pertaining to, + reaction): a chemical reaction that releases energy; a nonspontaneous and "downhill" process. Opposite of endergonic reaction.

exocarp (*ex-*, external to, + *-carp*, fruit): the outermost of the three layers of the pericarp (fruit wall) of angiosperms.

exocrine gland (*ex-*, external to, + *-crine*, to secrete): a gland that secretes its product into a duct that empties onto an external surface; such as a sweat gland. Opposite of endocrine gland.

exocytosis (*ex-*, external to, + *-cyt-*, cell, + *-sis*, the process of): a form of active transport in which cytoplasmic vesicles, containing waste materials, fuse with localized regions of the cell membrane so that the vesicle's contents can be expelled from the cell. Opposite of endocytosis.

exogenous (*ex-*, external to, + *-gen-*, origin, + *-ous*, pertaining to): produced or caused by external factors; such as the exogenous application of hormones on plants. Opposite of endogenous.

exon (*ex-*, external to, + *-on*, a particle): a region of DNA that is transcribed and expressed (as a protein); codes for the amino acid sequence of a specific polypeptide. Exons are separated by introns.

exoskeleton (*ex-*, external to, + *-skelet-*, a dried hard body): an external structure or covering secreted by the epidermis used for protection and the attachment of muscles; characteristic of arthropods and some mollusks. Opposite of endoskeleton.

exotoxin (*ex-*, external to, + *-tox-*, a poison, + *-in*, chemical substance): a toxin produced and secreted by intact bacteria; often an extremely potent poison. Opposite of endotoxin.

exponential growth: a pattern of growth characterized by a steeply climbing, almost vertical curve, representing unchecked population growth; growth of a population that repeatedly doubles its size. A somewhat theoretical pattern since natural populations cannot sustain such a growth rate.

extant (*extant-*, to stand out): living, present-day species. Opposite of extinct.

extensor (*ex-*, out, away from, + *-tend-*, to stretch, + *-or*, result of the act of): a muscle that extends a structure, such as a limb. Opposite of flexor.

extinct (*extinct-*, to be extinguished): the permanent loss of all individuals of a species. Opposite of extant.

extracellular matrix (*extra-*, outside of, + *-cell-*, a small room, + *-ar*, pertaining to, + *matr-*, mother): any substance normally found outside of cell membranes; such as blood plasma. See matrix.

extraembryonic membranes: four membranes that protect and support developing embryos of amniotes (mammals, birds, and reptiles). See amnion, chorion, allantois, and yolk sac.

eyespot: a collection of light sensitive pigments found in some unicelluar flagellated organisms such as *Euglena;* also known as a stigma.

F

F_1 (first filial) generation (*filia-*, daughter): the first (F_1) generation of offspring from a genetic cross.

F_2 (second filial) generation (*filia-*, daughter): the second (F_2) generation of offspring from a genetic cross between individuals from the same F_1 generation.

fascia—see adventitia

facilitated diffusion (transport) (*facilit-*, easy, + *diffus-*, to pour out, + *-ion*, process of): the spontaneous movement of molecules, bound to carrier proteins, across a cell membrane from a region of high concentration to a region of low concentration (down a concentration gradient). No source of chemical energy (ATP) is required.

facultative anerobe (*facult-*, faculty, + *-ive*, tending to, + *an-*, without, + *-aer-*, air + *-bi-*, life): an organism that can live in the absence of free molecular oxygen but usually does not.

FAD (Flavine Adenine Dinucleotide): an important coenzyme that serves as an electron acceptor in the Krebs cycle and other metabolic oxidation-reduction reactions. See electron acceptor.

fascicle (*fasci-*, a bundle, + *-cle*, little): any tight cluster or bundle; usually in reference to the arrangement of needle-like gymnosperm leaves, or a bundle of muscle cells.

fat—see adipose, lipid, and triglyceride

fatty acid: an organic molecule consisting of a long hydrocarbon chain bonded to an acidic carboxyl group (-COOH). Fatty acids are the basic units of fats, oils, and phospholipids.

fauna (*faun-*, god of the woods): all the animals of a particular area, or period of time.

fenestra, ***pl.*** **fenestrae** (*fenestr-*, opening): a microscopic opening in a capillary wall which permits movement of molecules to and from the capillary.

fermentation (*ferment-*, ferment, + *-ation*, the process of): anaerobic respiration that breaks down organic compounds (usually carbohydrates) to yield energy, carbon dioxide, and alcohol or lactic acid; less efficient (less ATP is made available) than aerobic respiration since an electron transport chain is not used.

ferredoxin (*ferr-*, iron, + redox, a contraction of reduction and oxidation, + *-in*, chemical substance): an iron containing protein that is a primary electron acceptor in photosynthesis.

fertilization (*fertil-*, fertile, + *-zation*, the process of): the fusion of two haploid cells (gametes) resulting in a diploid zygote; also know as syngamy.

fetus (*fet-*, pregnant, + *-us*, person): a developmental stage of an animal that takes place within an egg or body of its mother. In human development, the fetus stage occurs from the ninth week of development until birth (the last two trimesters of pregnancy).

fiber, fibril (*fib-*, fiber, thread): thread-like cellular structures; technically, a fiber is a minute strand of cytoplasmic material secreted by a cell and lying outside the cell. A fibril is a strand of cytoplasmic material (often protein) produced by a cell and lying within the cell. Plant fibers are elongated, tapering sclerenchyma cells, often lignified and dead, and used for mechanical support.

fibrin (*fib-*, fiber, + *-in*, chemical substance): a thread-like protein that forms blood clots.

fibrous roots (*fibr-*, fiber, + *-ous*, pertaining to, + roots): a relatively wide root system composed of many roots of approximate equal length; typical of grasses.

fiddlehead: a tightly coiled, immature fern frond, so named because of its resemblance to the scroll of a violin; also known as a crozier.

filament (*fil-*, thread, + *-ment*, means of): a thread-like structure or growth form; such as the elongated stalk of a stamen, or the thread-like thalli of certain algae and fungi.

fission (*fiss-*, to split, + *-ion*, the process of): a form of asexual reproduction in which the parent cell splits into two halves. Characteristic of bacteria, blue-green algae, and some yeasts. See binary fission.

fitness (evolutionary or genetic): a measure of the ability of an individual or population to survive natural selection based on genetic composition; also referred to as adaptive value. See natural selection.

flaccid (*flacc-*, limp, + *-id*, pertaining to): a condition in plants caused by a lack of water, resulting in a loss of turgidity; also referred to as wilted. Opposite of turgid.

flagellum, ***pl.*** **flagella** (*flagell-*, a whip, + *-um*, structure): a long whip-like projection on the surface of some cells, used for locomotion and feeding; structurally similar to cilia but much longer and fewer in number. See axoneme.

flavonoids (*flav-*, yellow, + *-oid*, resembling): a group of yellow colored pigments especially abundant in flower petals; associated with attracting insect pollinators by reflecting ultraviolet light.

flexor (*flex-*, to bend, + *-or*, result of the act of): a muscle that flexes or bends a structure such as a limb. Opposite of extensor.

flora (*flor-*, flower): all the vegetation in a particular area, or period of time.

floret (*flor-*, flower, + *-et*, little): a single flower in the inflorescence of grass plants or plants in the Compositae family; such as sunflowers.

floridean starch (*florid-*, flowery, + *-an*, pertaining to, + starch): a carbohydrate food reserve unique to red algae.

flower (*flor-*, flower): the reproductive structure of angiosperms. See complete flower.

fluid mosaic model: a current view describing the structure and function of cell membranes. Several types of proteins form a mosaic pattern by "floating" in a "liquid-like" bilayer of phospholipid molecules.

fluke (*floc-*, flat fish): a parasitic flatworm; found in the blood, liver, or lungs of many vertebrates.

foliose (*foli-*, leaf, + *-ose*, resembling): a growth form resembling leaves (flat, broad, and prostrate); common to certain lichens.

follicle (*follic-*, small bag): (1) a simple, dry fruit that opens along one side. (2) a small sac or cavity within glandular tissue that functions to secrete and sometimes store hormones. (3) a small sac of cells in the mammalian ovary that produces an egg and secretes estrogen.

food chain: a sequence of food transfered from producers to various levels of consumers and finally ending with decomposers.

food vacuole—see vacuole

food web: a complex system of all interconnected food chains in a community.

foramen (*foram-*, an opening): a small opening or hole through which blood vessels or nerves pass.

fossa (*foss-*, a pit, to dig up): any low pit or depression, as in bones, on which a muscle attaches to the bone.

fossil (*foss-*, a pit, to dig up, + *-il*, little): the remains of living beings preserved in rock.

fovea (*fov-*, a depression): a depression in the retina of the eye where the sharpest image is formed.

fragmentation (*frag-*, to break, + *-ment*, means of + *-ation*, process of): (1) a method of asexual reproduction involving an organism breaking into pieces with each piece capable of growing into a complete organism; common to some algae, fungi, and certain echinoderms. (2) the separation of a small segment of DNA from a chromosome often causing a mutation.

free energy: the energy available to do work as a result of an exothermic chemical reaction; symbolized G.

frond (*frond-*, leaf): a fern leaf.

fruit: a mature ovary or ovaries usually containing seeds; unique to angiosperms.

fruiting body: the reproductive structure of most fungi that produces spores. Opposite of the vegetative body (mycelium). See mycelium.

frustule (*frust-*, a piece, + *-ule*, little): the siliceous cell wall (shell) of diatoms; each half of a frustule is referred to as a valve.

fruticose (*frutic-*, shrub, + *-ose*, resembling): a shrub-like, branching growth form; common to certain lichens.

fucoxanthin (*fuc-*, a seaweed, + *-xanth-*, yellow, + *-in*, chemical substance): a yellow-brown accessory pigment found in brown algae and some protists.

functional group (*funct-*, to perform, + *-al*, pertaining to, + group): an arrangement of atoms that gives a similar chemical behavior to all molecules that attach to it; such as amino groups (NH_2) or hydroxyl groups (-OH).

fungus, *pl.* **fungi** (*fung-*, mushroom, + *-us*, thing): a heterotroph that feeds off of and as a result decomposes, dead or decaying organic material; such as a mushroom, mold, and smut.

funiculus (*funicul-*, little rope): (1) a stalk-like structure in an ovary that attaches the ovule to the placenta. (2) the column of white matter on the spinal cord.

fusiform cell (*fus-*, spindle, + *-form*, shape, + cell): any elongated cell tapered at the ends; such as certain types of xylem.

G

G_1 phase: the first gap phase in the cell cycle during which normal metabolic processes take place; follows cytokinesis. See cell cycle.

G_2 phase: the second gap phase in the cell cycle during which growth occurs; follows the S (DNA synthesis) phase. See cell cycle.

GA—see gibberellins

gametangium (*gamet-*, union, + *-angi-*, container, + *-um*, part, region): a structure producing gametes. It may be unicellular as in some algae or multicellular as in some mosses.

gamete (*gamet-*, union): a haploid sex cell that fuses with another to form a diploid zygote; such as sperm (male) and egg (female).

gametic meiosis (*gamet-*, union, + *-ic*, pertaining to, + *mei-*, reduction, + *-sis*, the process of): the production of gametes by meiosis; as in most animals, and the brown algae *Fucus*. Opposite of sporic meiosis.

gametophyte (*gamet-*, union, + *-phyt-*, plant): the haploid phase of a plant life cycle that produces haploid gametes by mitosis. Opposite of sporophyte. See alternation of generations.

ganglion, *pl.* **ganglia** (*gangli-*, a swelling): a knot-like mass of nerve cell bodies located outside of the Central Nervous System.

gap junction (gap, + *junct-*, to join, + *-ion*, process of): a protein-lined channel that serves as the primary molecular and electrical communications link between animal cells.

gastric (*gastr-*, stomach, + *-ic*, pertaining to): pertaining to the stomach, such as gastric glands.

gastrovascular cavity (*gastr-*, stomach, + *-vascul-*, a small vessel, + *-ar*, pertaining to, + cavity): a central cavity with just one opening that functions in digestion, respiration, and circulation; characteristic of primative invertebrates including sponges and cnidarians.

gastrula (*gastr-*, stomach, + *-ula*, little): an embryonic stage of animal development following the blastula stage, during which all three germ layers form. See germ layer.

gastrulation (*gastr-*, stomach, + *-ula*, little, + *-tion*, the process of): the formation of a gastrula from a blastula; characterized by an invagination of the cell layers to form a cap-like structure.

Gause's principle—see competitive exclusion principle
gel electrophoresis—see electrophoresis
gemma, *pl.* gemmae (*gemm-*, bud): a disk-like, asexual reproductive structure produced on the thallus of liverworts and some mosses.
gene (*gen-*, origin): the basic hereditary unit composed of a sequence of DNA nucleotides on a chromosome. One gene typically codes for one protein or RNA molecule.
gene amplification (*gen-*, origin, + *ampli-*, to increase, + *-ation*, the process of): a nuclear process in which extra copies of a specific gene are manufactured to increase the production of a specific molecule; such as the gene for ribosomal RNA in cells (oocytes) requiring rapid bursts of protein synthesis.
gene flow: the movement of genes from one population to another due to the migration of individuals. Such a flow often provides a significant source of genetic variation.
gene frequency: the relative occurrence of a particular gene in a population.
gene pool: all the genes of all the individuals in a population.
gene therapy (*gen-*, origin, + *therap-*, care, + *-y*, process of): the use of genetic engineering to treat or cure diseases caused by a faulty gene or the complete absence of a gene. This procedure is done by inserting a normal gene into an organism's genome.
generative cell (*gener-*, to produce, + *-ive*, tending to, + cell): a cell in angiosperm pollen grains that divides to produce two sperm cells.
genetic code (*gen-*, origin, + *-tic*, pertaining to, + code): sequences of nucleotide "triplets" (codons) in DNA that ultimately determine the amino acid sequence in proteins. May also be referred to as a triplet code. See codon.
genetic drift (*gen-*, origin, + *-tic*, pertaining to, + drift): a change in gene frequency due to random chance; usually associated with relatively small populations. Such a change often provides a significant source of genetic variation.
genetic engineering (*gen-*, origin, + *-tic*, pertaining to, + engineering): the manipulation and transfer of genes between chromosomes of the same species or between different species to increase crop yields, control harmful insects biologically, or produce more effective drugs. See recombinant DNA and gene therapy.
genetic equilibrium (*gen-*, origin, + *-tic*, pertaining to, + *equ-*, equal, + -libr-, balance, + *-ium*, part): a condition where the gene frequency of a population remains constant from generation to generation.
genetics (*gen-*, origin, + *-tic*, pertaining to): the study of heredity and heritable information.
genome (*gen-*, origin, + *-ome*, mass, group): the complete set of an organism's genes.
genotype (*gen-*, origin, + *-typ-*, form): the complete genetic makeup of an organism, whether or not it is expressed in the organism's phenotype. See phenotype.

geotropism (*ge-*, earth, + *-trop-*, to turn, + *-ism*, the process of): a direction of growth in response to gravity. Also known as gravitropism. Roots are positively geotropic and shoots are negatively geotropic.
germ cells (*germ-*, to sprout, + cells): gametes or the cells that give rise to gametes. Opposite of somatic cells.
germ layer (*germ-*, to sprout, + layer): one of three embryonic tissue layers (ectoderm, mesoderm, and endoderm) characteristic of most multicellular animals from which all tissues and organ systems arise.
germination (*germin-*, to sprout, + *-ation*, the process of): the initial growth and development by a plant from a seed or spore; a period of time from the uptake of water of a seed to the beginning of photosynthesis.
gestation (*gest-*, to carry, + *-ation*, the process of): the period of embryonic development from the time of conception to birth.
gibberellins (named after the fungal genus *Gibberella* from which they were first extracted): a group of hormones responsible for stem cell elongation, seed germination, and breaking bud dormancy; abbreviated GA.
gill: (1) a specialized respiratory organ of aquatic animals. (2) thin radially arranged lamella on the underside of the cap of certain Basidiomycetes ("mushrooms") on which spores are produced.
girdling (*girdl-*, to tie with a belt, + *-ing*, having the quality of): the removal of bark from a woody plant often resulting in the death of the plant; also known as ringing.
glial cells—see neuroglia cells
glomerulus (*glomerul-*, little ball, + *-us*, thing): a small round ball of capillaries contained within a Bowman's capsule of the kidney; the major site of blood filtration.
glucagon (*gluc-*, sweet, + *-agon*, to fight): a hormone produced by the alpha cells of the pancreas that raises blood sugar (glycogen) levels.
glucose (*gluc-*, sweet, + *-ose*, resembling): the most common 6-carbon monosaccharide ($C_6H_{12}O_6$) in plants; a common energy source.
glycerol (*glycer-*, sweet, sugar, + *-ol*, denoting an alcohol): one of the basic molecular components of fats and oils.
glycocalyx (*glyc-*, sweet, sugar, + *-caly-*, cup): the outer covering of animal cells made of oligosaccharide (the carbohydrate appendage of a glycoprotein molecule) lipids.
glycogen (*glyc-*, sweet, sugar, + *-gen*, to produce): the major type of carbohydrate stored in animal cells; composed of branching chains of glucose. Often referred to as animal starch.
glycolipid (*glyc-*, sweet, sugar, + *-lip-*, fat, + *-id*, pertaining to): a component of cell membranes composed of a hydrophilic carbohydrate "head" bonded to a hydrophobic lipid "tail." Such a molecule is said to be amphipathic. Glycolipids anchor surface proteins to the cell membrane. See amphipathic molecule.
glycolysis (*glyc-*, sweet, sugar, + *-lys-*, to dissolve): a metabolic pathway common to all organisms that splits glucose into pyruvate or lactic acid with a corresponding release of energy (ATP).

glycoprotein (*glyc-*, sweet, sugar, + *-prote-*, first, + *-in*, chemical substance): a protein covalently linked to a carbohydrate; most secretory proteins are glycoproteins.

glyoxysome (*glyc-*, sweet, sugar, + *-oxy-*, oxygen, + *-som*, body): a microbody containing digestive enzymes used to convert fat to sugar; found in the endosperm tissue of germinating seeds.

goblet cell: a mucus-secreting cell located in the epithelial lining of the digestive and respiratory tracts of vertebrates, so named because of its cup-like shape when viewed longitudinally.

Golgi complex (apparatus) (named after the Italian histologist, C. Golgi): a membrane-bound organelle that functions in processing, packaging, distributing, and secreting proteins.

gonad (*gon-*, seed): a gamete-producing organ; such as a testis or ovary.

gradualism (*grad-*, a step, + *-ule*, little, + *-ism*, process of): a theory that attributes evolutionary change in species to a slow and gradual accumulation of many subtle genetic changes over long periods of time. Opposite of punctuated equilibrium.

grain—see caryopsis

Gram-negative bacteria (named after the Danish physician, Hans Christian Gram): a group of bacteria whose cells do not take up crystal violet stain (when using the Gram method) due to the presence of a lipid bilayer surrounding the cell wall; appear pale red to pink.

Gram-positive bacteria (named after the Danish physician, Hans Christian Gram): a group of bacteria whose cells take up crystal violet stain (when using the Gram method) due to the presence of peptidoglycan in the cell wall; appear dark purple.

granulocyte (granular leukocyte) (*granul-*, little granules, + *-cyte*, cell): one of two classes of white blood cells that contain distinct cytoplasmic granules (lysosomes); involved with the inflammatory response of the immune system. Neutrophils, eosinophils, and basophils are granulocytes and are classified according to their staining properties. The other class of white blood cells is agranulocytes. See agranulocyte.

granum, *pl.* grana (*gran-*, granular, + *-um*, structure): stacked thylakoids (membrane-bound disks) within chloroplasts that contain alternate layers of chlorophyll, lipids and protein; the functional unit of photosynthesis.

gravitropism—see geotropism

gray matter: nervous tissue lacking myelinated nerve fibers and therefore appearing gray, as opposed to white matter which is myelinated and therefore white.

greenhouse effect: the warming of the earth's surface resulting from an accumulation of carbon dioxide in the atmosphere that absorbs infrared radiation (heat) and prevents it from being emitted back into space. The accumulation of carbon dioxide has been attributed to pollution and the burning of fossil fuels.

gross productivity: the total amount of chemical energy (in kilojoules) plants capture through the process of photosynthesis per area per unit time.

ground tissue (system): one of three fundamental tissue systems in plants that gives rise to pith and cortex. Also know as ground meristem. See also dermal tissue and vascular tissue.

growth regulator: a synthetic or naturally occurring organic compound active in controlling plant growth and development; such as ethylene.

growth ring: a layer of secondary xylem as seen in a stem cross section that usually corresponds to one season's growth.

guanine (after the South American Quechua word *huanu* meaning dung, + *-ine,* having the character of): one of four nitrogenous nucleotide bases that is a component of nucleic acids (DNA and RNA); complimentary to cytosine in DNA. It also occurs in the excrement (guano) of certain animals. See purine.

guard cells: two specialized plant epidermal cells that open and close the stoma (pore) as a result of changes in their turgor pressure.

guttation (*gutt-,* tear, + *-ation,* the process of): the exudation of water from specialized regions of leaves called hydathodes as a result of root pressure. Often mistaken for dew drops on leaves.

gymnosperm (*gymn-,* naked, + *-sperm,* seed): a plant producing seeds not enclosed within a fruit (mature ovary) but rather within a cone (strobilus). Also known as a cone-bearing plant.

gynoecium (*gyn-,* women, + *-eci-,* dwelling, + *-um,* structure): the female part of a flower; a collective term for all the carpels. Opposite of androecium.

H

habit (*habit-,* character, condition): the characteristic behavior, form, or mode of growth of an organism; such as hairy, tall, climbing, nocturnal, etc.

habitat (*habitat-,* to live in): the natural environment where an organism is usually found; an organism's "address."

habituation (*habit-,* character, condition, + *-ation,* the process of): a type of learning in which continued exposure to the same relatively unimportant environmental stimulus results in a diminished response.

halophyte (*hal-,* salt, + *-phyt-,* plant): a plant that typically grows in a saline environment.

haploid (*hapl-,* half, + *-oid,* resembling): a normal condition in which a cell or organism has one set of parental chromosomes per nucleus; characteristic of gametes and plant spores; symbolized n. Opposite of diploid.

haplostele (*hapl-,* half, + *-stel-,* a pillar): the simplest type of protostele in which a central core of xylem is surrounded by phloem.

Hardy-Weinberg equilibrium (named after the English mathematician, G. Hardy and the German physician, W. Weinberg): a fundamental genetic and evolutionary principle that describes the conditions needed for the frequencies of genes in a population to stay the same; a measure of genetic stability or instability in a population's gene pool.

Hartig net (named after the German botanist, H. J. Hartig): a network of fungal hyphae interwoven with the cortical root cells of certain gymnosperm species.

haustorium (*haust-,* to drink, + *-ium,* region): a hypha or root used by parasites to penetrate and absorb nutrients from host tissues.

head: a type of inflorescence in which the flowers are clustered together forming a radial disk; such as a sunflower.

heartwood: secondary xylem centrally located in stems that does not transport water due to the accumulation of resins, oils, and gums; often dark in color and surrounded by sapwood. See sapwood.

helper T cell: a modified T cell that aids other T cells to respond to antigens or secrete interleukins, or helps B cells produce antibodies.

heme (haem) group (*hem-,* blood): a nonprotein (prosthetic) molecule containing iron; an essential component of cytochrome and hemoglobin molecules.

hemicellulose (*hemi-,* one half, + *-cellul-,* a small cell, + *-ose,* resembling): a highly soluble form of cellulose found in cell walls.

hemoglobin (*hem-,* blood, + *-glob-,* sphere + *-in,* a chemical substance): a red, iron-containing protein in red blood cells that combines with and transports oxygen and carbon dioxide; also aids in regulating blood pH.

hemolysis (*hem-,* blood, + *-lys,* to dissolve): the destruction of red blood cells.

hemophilia (*hem-,* blood, + *-phil-,* to love, + *-ia,* condition of): a genetic disorder resulting from an abnormal sex-linked recessive gene; characterized by the inability to clot blood.

hepatic (*hepat-,* liver, + *-ic,* pertaining to): pertaining to the liver; such as the hepatic artery.

herbaceous (*herb-,* grass, + *-aceous,* pertaining to): pertaining to soft green plants with little or no woody tissue; typical of most annuals and biennials.

herbicide (*herb-,* grass, + *-cide,* to kill): a compound that will kill or at least inhibit plant growth.

herbivore (*herb-,* plant, + *-vor-,* to eat): an animal that eats plant material.

hermaphrodite—see bisexual

herpes (*herpe-,* to creep):**—see type I herpes, and type II herpes**

heterochromatin (*heter-,* different, + *-chromat-,* color, + *-in,* chemical substance): compacted and tightly coiled chromatin in an inactive (nontranscribed) state; stains more intense than euchromatin. See euchromatin.

heterocyst (*heter-,* different, + *-cyst,* bag): a thick-walled, transparent cell in which nitrogen fixation occurs; found in filamentous blue-green algae.

heterodont (*heter-,* different, + *-odont,* tooth): an organism with teeth differentiated into canines, incisors, and molars. Opposite of homodont.

heterogamy—see anisogamy

heteromorphic (*heter-,* different, + *-morph-,* shape, + *-ic,* pertaining to): morphologically different; usually used to describe a plant life history in which the diploid and haploid generations differ in form. Opposite of isomorphic.

heterosporous (*heter-,* different, + *-spor-,* spore, + *-ous,* pertaining to): producing two types of spores: smaller microspores that give rise to male gametophytes, and larger megaspores that give rise to female gametophytes.

heterotroph (*heter-,* different, + *-troph,* nutrition): an organism unable to synthesize organic material and so must ingest preformed organic compounds; such as fungi and animals. Opposite of autotroph.

heterozygote advantage (*heter-*, different, + *-zyg-*, a pair, + advantage): a genetic situation in which the heterozygous condition (Aa) provides a greater degree of fitness than the homozygous condition (AA or aa); sickle-cell anemia is a well documented example.

heterozygous (*heter-*, different, + *-zyg-*, a pair, + *-ous,* pertaining to): having two different alleles for a specific genetic trait; such as Aa. Opposite of homozygous.

hibernation (*hibern-*, winter, + *-ation,* the process of): a temporary dormant stage characterized by decreased metabolic activity allowing some animals to survive long periods of cold and diminished food supply.

High Density Lipoproteins (HDL): plasma proteins that transport cholesterol from body tissues to the liver for destruction; so-called "good" lipoproteins because they lower cholesterol levels in the blood.

Hill reaction (named after the British chemist, R. Hill): a chemical reaction showing that illuminated choloroplasts isolated from plants can reduce various compounds including ferrocyanide and release oxygen as a by-product.

hilum (hilus) (*hil-*, a trifle, + *-um,* structure): (1) the central portion of a starch grain around which starch molecules are concentrically layed down. (2) the scar left on the seed after detachment from the funiculus. (3) the opening where ducts, nerves, and blood vessels enter or leave an organ.

histamine (*hist-*, tissue, + *-amine,* resinous gum): a substance produced by mast cells in response to specific antigens causing blood vessels to dilate as well as change their permeability; involved with allergic and inflammatory responses.

histogenesis (*hist-*, tissue, + *-gen-*, origin, + *-sis,* process of): the formation and development of tissue.

histone (*hist-*, tissue, + *-on,* a particle): a relatively small protein found in the nucleus that binds to the DNA and plays a role in the folding and condensation of DNA into chromatin.

HIV—see Human Immunodeficiency Virus

holdfast: a root-like organ used by certain algae to attach to rocks, but that does not absorb any water or nutrients.

holozoic (*hol-*, whole, + *-zo-*, animal, + *-ic,* pertaining to): a form of nutrition involving the ingestion of solid or liquid food particles; common to most animals including man.

homeostasis (*hom-*, same, + *-sta-*, to control, + *-sis,* process of): the maintenance of a balanced internal environment of the cell or body by means of a self-regulatory mechanism; such as blood pressure or pH. The maintenance of equilibrium between organism and environment.

homeotherm—see endotherm

hominid (*homo-*, man, + *-id,* member or offspring of): a member of the family Hominidae that includes humans and all closely related primates both extinct and living.

hominoid (*homo-*, man, + *-oid,* resembling): a group of animals including the great apes and man.

homodont (*homo-*, same, + *-odont,* tooth): an organism with all teeth uniform in size and shape. Opposite of heterodont.

homologous chromosomes (homologues) (*homolog-*, agreeing, + *-ous,* pertaining to, + *chrom-*, color, + *-some,* body): pairs of morphologically similar chromosomes that possess genes for the same traits at corresponding loci; each member of the pair is derived from a different parent.

homologous structures (*homolog-*, agreeing, + *-ous,* pertaining to, + structures): structures in different organisms that are similar in form because of a common origin; such as the forelimb of vertebrates.

homosporous (*hom-*, same, + *-spor-*, spore, + *-ous,* pertaining to): producing one type of spore. Opposite of heterosporous.

homozygous (*hom-*, same, + *-zyg-*, a pair, + *-ous,* pertaining to): having two identical alleles for a specific genetic trait; such as AA or aa. Also referred to as pure breeding. Opposite of heterozygous.

hormogonium (*horm-*, chain, + *-gon-*, reproduction, + *-ium,* region): a segment of a blue-green algal filament capable of growing into a new filament; a means of asexual reproduction.

hormone (*hormone,* to excite): a naturally occurring organic compound produced in one part of the organism and transported to another part, where it has, even in minute amounts, a pronounced effect on growth and development.

host: (1) an organism to which a parasite attaches and lives off of. (2) an organism that receives transplanted or grafted tissue.

Human Immunodeficiency Virus (HIV): the retrovirus responsible for AIDS. See Acquired Immune Deficiency Syndrome.

humoral immunity (*humor-*, fluid, + *-al,* pertaining to, + *immuni-*, free, + *-ty,* state of being): immunity against bacteria and viruses mediated by antibodies carried in the blood and lymph (bodily fluids formerly referred to as "humors").

humus (*hum-*, soil, + *-us,* thing): the organic, usually uppermost, layer of soil.

hybrid (*hybrid,* mixed offspring): the offspring from a cross of genetically different parents, varieties, or species.

hybrid inviability: a postzygotic isolating mechanism in which the embryos of interspecific hybrids do not develop and abort.

hybrid sterilty: a postzygotic isolating mechanism in which hybrid organisms cannnot produce viable gametes.

hybrid vigor: increased genetic superiority in the hybrid offspring of two different parents due to an increased number of heterozygous alleles.

hybridoma (a contraction of hybrid and melanoma): a hybrid cell produced by the fusion of a cancer (melanoma) cell and a normal lymphocyte; used in the manufacture of monoclonal antibodies. See monoclonal antibody.

hydathode (*hydr-*, water, + *-ode,* way): a specialized region of leaves at the ends of veins from which water is exuded during guttation. See guttation.

hydrogen bond: a relatively weak chemical attraction formed between a hydrogen atom possessing a partial positive charge and an atom (usually oxygen or nitrogen) possessing a partial negative charge.

hydrolysis (*hydr-*, water, + *-lys-*, to dissolve): a common catabolic reaction in which one large molecule is split into two smaller ones by the addition of a water molecule (H^+ and OH^-). Hydrolytic enzymes are usually involved. Opposite of dehydration synthesis.

hydrophilic (*hydr-*, water, + *-phil-*, to love, + *-ic,* pertaining to): having a chemical affinity for water; associated with polar molecules or polar regions of larger molecules. Hydrophilic compounds are soluble in water. Opposite of hydrophobic.

hydrophobic (*hydr-*, water, + *-phob-*, fear, + *-ic,* pertaining to): having no chemical affinity for water; associated with nonpolar molecules or nonpolar regions of larger molecules. Hydrophobic compounds are insoluble in water. Opposite of hydrophilic.

hydrophyte (*hydr-*, water, + *-phyt-*, plant): a plant that typically grows in water or in very moist environments.

hydroponics (*hydr-*, water, + *-pon-*, exertion, + *-ic,* pertaining to): a method of growing plants in a liquid nutrient medium without soil.

hydrostatic skeleton (*hydr-*, water, + *-stat-*, to control, + *-ic,* pertaining to + skeleton): a support system composed of pressurized fluid-filled cavities; common to many invertebrates.

hydroxyl group (a contraction of hydrogen and oxygen, + *-yl,* chemical radical): -OH; a common functional group or ion. Compounds with this group include acids, bases, alcohols, phenols, and are readily soluble in water.

hymenium (*hymen-*, a membrane, + *-ium,* region): the fertile or spore-producing layer of tissue in certain fungi; such as Basidiomycetes and Ascomyetes.

hypaxial (*hypo-*, below, + *-ax-*, axis, + *-ial,* pertaining to): usually pertaining to muscles located below or ventral to the vertebral column. Opposite of epaxial.

hyperplasia (*hyper-*, above, + *-plas-*, development, + *-ia,* condition of): a type of growth, either normal or abnormal, resulting from an increase in the number of cells rather than from an increase in cell size. The growth of skin is hyperplasic. Opposite of hypertrophy.

hypertonic (solution) (*hyper-*, above, + *-ton-*, condition of, + *-ic,* pertaining to): referring to a solution that has a greater concentration of solute than of water compared with another solution. Water will move across a selectively permeable membrane from a hypotonic solution to a hypertonic solution. Also referred to as hyperosmotic. See hypotonic.

hypertrophy (*hyper-*, above, + *-troph-*, nutrition, + *-y,* state of): a type of growth, either normal or abnormal, resulting from an increase in cell size rather than from an increase in cell number. Skeletal muscle growth is hypertrophic. Opposite of hyperplasia.

hypha, ***pl.*** **hyphae** (*hypha-*, something woven): a single thread-like filament of the vegetative body (mycelium) of most fungi; collectively all hyphae are referred to as mycelium.

hypocotyl (*hypo-*, below, + *-cotyle-*, cup): the portion of an embryo or seedling between the cotyledons and the radicle.

hypodermis (*hypo-*, below, + *-derm-*, skin): one or more layers of cells immediately below the epidermis of plants.

hypothalamus (*hypo-*, below, + *-thalam-*, inner chamber, + *-us*, thing): a region of the vertebrate brain below the thalamus that functions in regulating homeostasis, emotions, the autonomic nervous system, and the pituitary gland.

hypotheca (*hypo-*, below, + *-thec-*, case): the innner, smaller half of the silicious portion of a diatom cell wall. Opposite of epitheca.

hypothesis (*hypo-*, below, + *-the-*, to put, + *-sis*, the process of): a tentative explanation to a specific problem based on the accumulation of information that must be scientifically tested to be valid. If a hypothesis is not supported experimentally it must be discarded or modified. See scientific method.

hypotonic (solution) (*hypo-*, below, + *-ton-*, condition of, + *-ic*, pertaining to): referring to a solution that has a greater concentration of water than of solute compared with another solution. Water will move across a selectively permeable membrane from a hypotonic solution to a hypertonic solution. Also referred to as hyposmotic. See hypertonic.

I

IAA—see Indol-Acetic Acid

ileum (*ileum*, groin, small intestine): the last (lower) portion of the small intestine in vertebrates extending from the jejunum to the large intestine; where bile salts and vitamin B_{12} are absorbed.

imbibition (*imbib-*, to drink, + *-tion*, the process of): the physical uptake of water into a very dry tissue or organ; such as a seed.

immigration (*im-*, in, + *-migr-*, to move, + *-ation*, the process of): the process of moving into a new location to live. Opposite of emigration.

immune response (*immuni-*, free, + response): a specific bodily defensive mechanism in response to antigens involving the production of B lymphocytes (which produce antibodies) and T lymphocytes (which are responsible for cell-mediated immunity).

immunoglobulin (*immuni-*, free, + *-globul-*, round, + *-in*, chemical substance): an antibody; a complex globular protein molecule produced in response to the presence of a foreign substance (antigen) in the body and having the ability to react against the antigen.

imperfect flower (*imperfect-*, unfinished, + flower): a flower lacking either stamens or pistils. Also known as unisexual. Opposite of perfect flower.

imperfect fungi (*imperfect-*, unfinished, + *fung-*, mushroom): fungi in which sexual reproduction does not occur or has not been observed; all such fungi reproduce asexually and are often placed into the division Deuteromycota. Also referred to as ''Fungi Imperfecti.''

implantation (*im-*, in, + *-plant-*, to plant, + *-ation*, the process of): (1) the process of attaching the developing embryo to the lining of the uterus. (2) grafting tissue or inserting an organ onto a new location.

imprinting: a type of rapid learning characteristic of young birds and mammals in which the young will recognize and form a strong social attachment to the first moving object they see.

incomplete dominance: a hereditary mechanism in which both alleles of a heterozygous pair are expressed, thus producing a phenotype intermediate between the two alleles.

incomplete flower: a flower lacking any of the four floral parts; such as sepals, petals, stamens, or pistils (carpels). Opposite of complete flower.

indehiscent (*in-*, not, + *-dehisc-*, to split, + *-ent*, having the quality of): usually in reference to fruit not splitting open along a distinct line or suture; such as an apple. Opposite of dehiscent.

independent assortment (Mendel's second law): a fundamental principle of genetics in which the alleles of unlinked genes (those on separate chromosome pairs) assort independently of each other during meiosis. Genes located on separate pairs of chromosomes are inherited independently of each other.

indeterminate cleavage (*in-*, not, + *-determin-*, to end completely, + *-ate*, characterized by having, + *cleav-*, to divide, + *-age*, collection of): a type of embryonic development in animals in which the fate of each embryonic cell is capable of developing into any body part; usually a radial pattern of cleavage and characteristic of deuterostomes. See radial cleavage.

indeterminate growth (*in-*, not, + *-determin-*, to end completely, + *-ate*, characterized by having, + growth): a growth pattern characteristic of plants (and few animals) in which growth is unrestricted; usually associated with meristematic activity. Opposite of determinate growth.

indigenous (*indigen-*, native, + *-ous*, pertaining to): pertaining to an organism that is native to a specific area; not introduced. Opposite of introduced.

Indole-Acetic Acid (IAA): a naturally occurring auxin responsible for controlling cell elongation, fruit development, and other processes in plants. See auxins.

induced fit: a change in the 3–D shape of an enzyme's active site in order to enhance the "fit" between the active site and substrate molecule. The change in shape is caused by the entry of the substrate molecule into the enzyme.

induction (*induce-*, to lead in, + *-tion*, the process of): a developmental process in which one group of embryonic cells or tissues influence the development of other cells or tissues.

indusium, ***pl.*** **indusia** (*indus-*, garment, + *-ium*, region): a thin layer of epidermal tissue covering a fern sorus.

inferior ovary (*infer-*, below, + *ov-*, an egg, + *-ary*, place for): an ovary situated below the stamens, petals, and sepals. Also referred to as an epigynous ovary. Opposite of superior ovary.

inflammation (*in-*, into, + *-flamm-*, flame, + *-ation*, the process of): a reddening and swelling of tissue caused by the dilation of blood vessels in response to an infection or injury; also referred to as an inflammatory response.

inflorescence (*infloresc-*, begin to bloom, + *-ence*, the condition of): (1) a cluster of individual flowers in a distinct arrangement. (2) the specific arrangement of flowers on a stem.

ingestion (*in-*, into, + *-gest-*, to carry, + *-ion*, process of): a form of heterotrophic nutrition in which organisms are eaten whole or in pieces.

initiator codon: a codon found in messenger RNA that marks where protein synthesis begins. Opposite of terminator codon.

inoculum (*inocul-*, engrafted, + *-um*, structure): a small amount of living cells or tissue used to start a culture *in vitro*. See *in vitro*.

insectivorous (*insect-*, cut into, + *-vor-*, to eat, + *-ous*, pertaining to): (1) pertaining to plants with stems or leaves modified for trapping insects. These plants obtain inorganic nutrients from the insects. (2) pertaining to an animal that eats primarily insects; such as some birds.

insertion (of a muscle): the end of a muscle that attaches to a movable structure, usually a bone. Opposite of origin.

insulin (*insul-*, island, + *-in*, chemical substance): a hormone produced and secreted by the beta cells in the pancreas that lowers glucose levels in the blood by promoting the uptake of glucose by body cells; also stimulates lipid and protein synthesis.

integral proteins (*integ-*, whole, + *-al*, pertaining to, + *prote-*, first, + *-in*, chemical substance): protein molecules found embedded in and often spanning the entire width of a cell's plasma membrane; also referred to as transmembrane proteins. These proteins are associated with several functions including active transport, cell recognition, and cell to cell adhesion.

integument (*integ-*, a covering, + *-ment*, condition of): (1) an external layer or covering of cells; such as the skin. (2) the outermost layer of the ovule that develops into the seed coat.

intercalary growth (*intercal-*, to insert between, + *-ary*, place for, + growth): growth restricted to a region(s) between the apex and the base of the plant; such as at nodes. Opposite of apical growth.

intercalated disks (*intercal-*, to insert between, + *-ate*, characterized by having, + disks): specialized junctions that hold adjacent cardiac muscle cells together and that appear as dense bands at right angles to the muscle striations; a characteristic of vertebrate cardiac muscle.

intercellular (*inter-*, between, + *-cellul-*, a small cell, +, *-ar*, pertaining to): referring to any material or object produced or found between cells.

intercourse—see copulation

interferons (*inter-*, between, + *-fer-*, to carry, + *-on*, a particle): a group of naturally occurring proteins of the immune system produced by virus-infected cells possessing antiviral properties and capable of assisting other cells to resist viral infection. Several types of interferon have been identified and designated α–, β–, λ–.

interleukins (I, II, III) (*inter-*, between, + *-leuk-*, white, + *-in*, chemical substance): chemical substances produced by macrophages and T lymphocytes that function as metabolic regulators (cytokines) of themselves and of neighboring cells. Several important functions related to the immune system have been linked to interleukins including regulating body temperature and maintaining T and B cell populations.

internode (*inter-*, between, + *-nod-*, knot): the region of a stem between two successive nodes or joints; a region of the plant where leaves are attached.

interphase (*inter-*, between, + *-phase*, a stage): the longest phase of the cell cycle (including G_1, S, G_2) during which the chromosomes are not condensed (chromatin) and are actively involved with protein synthesis. See cell cycle.

interstitial (*inter-*, between, + *-stit-*, to stand, + *-ial*, pertaining to): (1) cells located in the fluid-filled spaces between other cells and organs; such as the testosterone secreting cells located between the seminiferous tubules of the testes. (2) growth resulting from an increase in the number of cells and amount of matrix within tissue; such as the growth of cartilage. Opposite of appositional growth.

intertidal zone—see littoral zone

intracellular (*intra-*, within, + *-cellul-*, a small cell, +, *-ar*, pertaining to): referring to any material or object produced or found within cells; such as nuclei. Opposite of extracellular.

intrinsic proteins (*intrins-*, internally, + *-ic*, pertaining to, + *prote-*, first, + *-in*, chemical substance): protein molecules having all or part of their molecular structure embedded in the plasma membrane. See integral proteins.

intron (*intra-*, within, + *-on*, a particle): a segment of DNA (or RNA) that does not code for protein; occuring between regions of DNA (or RNA) that do code for specific proteins (exons). The precise function of introns is not clearly understood; however, one hypothesis indicates introns may facilitate genetic recombination between exons during meiosis. See exon.

invagination (*in-*, into, + *-vagin-*, sheath, + *-ation*, the process of): the infolding of a layer of cells into another layer forming a cup-like depression; usually in reference to animal embryology and gastrulation. See gastrulation.

inversion (chromosomal) (*invert-*, to invert, + *-ion*, process of): a chromosomal abnormality in which a segment of a chromosome is broken off and reattached in a reverse orientation to the same chromosome; may be caused by a mutagen or an error in meiosis.

invertase (*invert-*, to invert, + *-ase*, enzyme): an enzyme that breaks sucrose down into glucose and fructose.

in vitro (in, + *vitro*, glass): in reference to experimenting with living cells that have been removed from their natural environment; such as *in vitro* fertilization in which an egg is fertilized outside the body.

in vivo (in, + *viv-*, living): in reference to naturally occurring biological events; such as the development of a fertilized egg within the mother.

involuntary muscle (*in-*, not, + *-volunt-*, of one's free will, + *-ary*, pertaining to, + muscle): usually in reference to the smooth muscle lining the digestive tract that is not subject to conscious control. Opposite of voluntary muscle.

ion (*ion*, to wonder): an atom or group of atoms possessing a net positive (cation) or net negative (anion) electrical charge. See anion and cation.

ionic bond (*ion-*, to wonder, + *-ic*, pertaining to, + bond): a relatively strong chemical bond formed as a result of the attraction between oppositely charged ions.

irregular flower: a flower that has bilateral symmetry. Also referred to as a zygomorphic flower. Opposite of regular flower. See bilateral symmetry.

irritability (*irritab-*, to provoke, + *-ity,* the state of): a fundamental property of life involving the ability of an organism to respond to an environmental stimulus, either positively or negatively.

islets of Langerhans (named after the German anatomist, P. Langerhans): distinct groups of endocrine cells within the pancreas in which insulin and glucagon are produced. See insulin and glucagon.

isogamy (*iso-*, same, + *-gam-*, union, + *-y,* state of): a type of sexual reproduction in which the gametes are morphologically identical; as in certain algae and fungi. Opposite of oogamy.

isomers (*iso-*, same, + *-mer,* a part of): organic compounds that have identical chemical formulas but different structural formulas (arrangements); such as glucose and fructose which are both 6-carbon sugars ($C_6H_{12}O_6$).

isomorphic (*iso-*, same, + *-morph-*, shape, + *-ic,* pertaining to): morphologically identical; used to describe a plant life history in which the diploid and haploid generations are identical in form. Opposite of heteromorphic.

isotonic (*iso-*, same, + *-ton-*, condition of + *-ic,* pertaining to): referring to two solutions that have identical concentrations of solvent and solute molecules. If separated by a selectively permeable membrane, no net flow of water will occur between two isotonic solutions. Also referred to as isosmotic.

isotope (*iso-*, same, + *-top-*, position): an alternate form of an element with the normal number of electrons and protons but a different number of neutrons and therefore a different molecular weight. Many isotopes are chemically unstable and radioactive.

J

jejunum (*jejun-*, empty, + *-um,* structure): the second (middle) portion of the small intestine in vertebrates located between the duodenum and the ileum and is responsible for absorbing organic nutrients; so named because it is often empty upon death.

joule (named after the British physicist, J.P. Joule): a unit of energy that has replaced the obsolete calorie (Calorie). 4.2 kilojoules equals approximately 1 Calorie. See calorie.

juvenile hormone (neotinin): a hormone common to arthropods that prevents metamorphosis, thereby maintaining larval or nymphal characteristics.

K

K-selection (K-strategist) (from the K value in the logistic equation): a survival strategy characterized by populations whose members mature relatively slowly, and produce relatively few offspring who receive abundant parental care. Most of the offspring survive to reproduce; such as humans. Opposite of r-selection. See logistic growth (logistic equation).

karyogamy (*kary-*, nut, nucleus, + *-gam-*, union, + *-y,* state of): the fusion of two gamete nuclei, usually resulting in a diploid condition.

karyokinesis—see mitosis

karyotype (*kary-*, nut, nucleus, + *-typ-*, form): a method of arranging and displaying all the chromosomes of an organism according to number, size, and shape; usually done photographically.

kelp: a common name for any large marine algae.

keratin (*kerat-*, horn, + *-in*, chemical substance): a hard, insoluble protein abundant in the skin of many vertebrates and in the derivatives of the skin; such as hair, horns, hooves, and claws.

kilocalorie—see calorie

kinetic energy (*kine-*, movement, + *-tic*, pertaining to): the energy of motion; measured in Joules. Opposite of potential energy.

kinetin (*kine-*, movement, + *-in*, chemical substance): a synthetic, purine-based cytokinin. See cytokinins.

kinetochore (*kine-*, movement, + *-chor-*, chorus): a disk-shaped protein structure on the centromere of chromosomes to which spindle fibers are attached during mitosis and meiosis. See centromere.

kinetosome—see basal body

Krebs cycle (named after the German biochemist, H. Krebs): the second stage of cellular respiration (between glycolysis and electron transport system) that completes the breakdown of glucose into carbon dioxide; occurs in the matrix of mitochondria and in the cytoplasm of some bacteria. Also referred to as the citric acid cycle or TCA (tricarboxylic acid) cycle.

L

labium, *pl.* **labia** (*labi-*, lip, + *-um*, structure): a fleshy lip or lip-like structure surrounding an opening; such as the paired appendages surrounding the mouth of certain insects, or the lip-like borders of the vulva in humans.

labyrinth (*labyrinth*, a tortuous passage): any complex network of interconnecting cavities or canals; such as the spiral canal of the inner ear of many vertebrates.

lactation (*lact-*, milk, + *-ation*, the process of): (1) the process of producing and secreting milk by mammary glands. (2) the period of time mammalian offspring are breast-fed.

lacteal (*lact-*, milk, + *-eal*, pertaining to): (1) a lymph vessel in the intestinal villi of mammals responsible for absorbing the components of lipid digestion. (2) relating to the production of milk.

lacuna, *pl.* **lacunae** (*lacun-*, small pit, cavity): any space or small cavity, as in bone where the metabolically active bone cells are located.

Lamarckism (named after the French naturalist, J. Lamarck): the evolutionary theory stating that traits acquired in an organism's lifetime are directly inherited by the offspring. Lamarck's theory was eventually disproven by A. Weismann.

lamella (*lamell-*, small thin plate): any thin, plate-like structure, such as the thin structures found in most gills.

lamina (*lamin-,* small thin plate)—**see blade**

laminarian (after the genus of brown algae, *Laminaria,* + -in, chemical substance): a storage polysaccharide unique to brown algae; so named because it is often extracted from the genus *Laminaria.*

larva, *pl.* larvae (*larva,* a ghost): an immature, free-living stage in the life cycle of some animals that is often distinctly different from the adult; such as a caterpillar which is the larval stage in the life cycle of a butterfly.

lateral (*later-,* side, + *-al,* pertaining to): pertaining to the side of an organ or organism; such as a lateral meristem.

lateral meristem (*later-,* side, + *-al,* pertaining to, + *merist-,* divisible): one or more layers of actively dividing plant cells that give rise to secondary tissues; located lateral to the longitudinal axis of the plant. See cork cambium and vascular cambium.

lateral root (*later-,* side, + *-al,* pertaining to, + root): a secondary or branch root that originates from the pericycle.

law: a theory that is virtually infallible under precisely controlled conditions; such as the law of independent assortment.

law of independent assortment—see independent assortment

law of segregation—see segregation

leaf gap: an interruption in the vascular tissue of stems; usually associated with a leaf trace.

leaflet: an individual unit of a compound leaf.

leaf primordium (leaf, + *primord-,* first, + *-ium,* region): an outgrowth from an apical meristem that eventually develops into a leaf.

leaf scar: a mark left on the stem after a leaf has fallen; often covered by a layer of suberin.

leaf trace: a strand of vascular tissue diverging from the stem stele and going into a leaf.

legume: (1) a member of the legume family; such as peas and beans. (2) a simple, dry fruit that splits along two sutures when mature.

lemma (*lemma,* husk): a small bract enclosing the flower in a spikelet of grass.

lenticel (*lenticel,* a small window): a small crack in the bark of woody plants that is used for gaseous exchange; also referred to as a ventilating pore.

leptotene (*lept-,* thin, delicate, + *-ten-,* to hold): the first stage in prophase I of meiosis during which the chromatin threads become visible and the centromeres develop along the chromosomes.

leucoplast (leukoplast) (*leuc-,* white, + *-plast,* membrane): a colorless plastid specialized for starch synthesis and storage. A precursor to all other types of plastids. See plastid.

leukemia (*leuc-,* white, + *-emia,* condition of the blood): a malignant form of cancer characterized by the uncontrolled proliferation and abnormal development of white blood cells in the blood and bone marrow.

leukocyte (leucocyte) (*leuk-,* white, + *-cyte,* cell): a white blood cell that functions in immunity; also referred to as white corpuscle. See granulocyte and agranulocyte.

Leydig's cells (named after the German anatomist, F. von Leydig): testosterone producing cells located in the interstitial tissue of testicles.

lichen: a composite organism made up of an algal partner (usually blue-green) and a fungal partner (usually an Ascomycete) living in a symbiotic or parasitic relationship.

ligament (*ligament-*, bandage): a tough sheet or band of dense connective tissue that joins bone to bone or supports organs within the abdominal cavity.

light dependent reaction: the first stage of photosynthesis that occurs on the thylakoid membranes of chloroplasts and that converts light energy into chemical energy (ATP and NADPH). See photolysis.

light independent reactions: the second stage of photosynthesis in which carbon dioxide is fixed (reduced) into organic compounds like glucose using energy from the first stage (light dependent reactions) of photosynthesis; these reactions can occur whether or not light is present. Also referred to as dark reactions or carbon fixation.

lignin (*lign-*, wood, + *-in*, chemical substance): an organic polymer associated with cellulose in some secondary cell walls and used to strengthen and harden the walls.

ligule (*ligul-*, small tongue): a small leaf-like outgrowth at the base of grass leaves and the leaves of certain lycopods.

limbic system (*limb-*, border, + *-ic*, pertaining to, + system): a complex network of neurons that forms a loop around the inside of the brain; responsible for emotions and short-term memory.

limiting factor: any environmental factor that prevents an organism from living in a specific habitat or restricts its population size. Such factors include food, temperature, light, and pH.

linkage (linked genes): the general tendency for one or more genes to be inherited together because they are located on the same chromosome.

lipase (*lip-*, fat, + *-ase*, enzyme): an enzyme secreted by the pancreas that digests lipids.

lipids (*lip-*, fat, + *-id*, pertaining to): a group of organic compounds that are insoluble in water (hydrophobic) including oils, fats, phospholipids, waxes, and steroid hormones. Some lipids function as integral components of cell membranes and as a long-term storage form of energy.

lipoprotein (*lip-*, fat, + *-prote-*, first, + *-in*, chemical substance): a molecule composed of a protein and a lipid that transports lipids between various regions of the body. Cholesterol travels in the blood in two forms: low-density lipoproteins (LDL), and high-density lipoproteins (HDL).

liposome (*lip-*, fat, + *-som-*, body): a spherical shell formed under laboratory conditions when phospholipids are mixed in an aqueous solution; important as a model for ancestral cell types. Also known as a protobiont.

littoral zone (*littor-*, seashore, + *-al*, pertaining to): a region of the ocean and shore between low and high tide; also known as intertidal zone.

locule (*locul-*, small chamber): any cavity, chamber, or compartment found in plants; such as a sporangium locule or ovarian locule.

locus, ***pl.*** **loci** (*locus,* place): the specific location of a gene or allele on a chromosome.

logistic growth (logistic equation): a mathematical model for population growth in which growth is rapid when the population is small, but as the population approaches the carrying capacity of the environment, the rate begins to slow. Eventually the rate of growth stabilizes and may decline as the population reaches its maximum size. An S-shaped growth curve represents this type of growth pattern. See carrying capacity. The logistic equation is:

$$\frac{dN}{dt} = rN\left(\frac{K - N}{K}\right)$$

long-day plants: plants that flower in response to increasing day lengths (usually greater than 12 hours) and a corresponding decrease in night length. Opposite of short-day plants.

longitudinal section: a section running along the length of a cylindrical object.

loop of Henle (named after the German anatomist, F. Henle): the U-shaped portion of the nephron that extends from the cortex down into the medulla of the mammalian and bird kidney; the primary site of water and salt reabsorption.

Low Density Lipoproteins (LDL): plasma proteins that transport cholesterol from the liver to body tissues; so-called "bad" lipoproteins because they increase cholesterol levels in the blood.

lumen (*lumen,* light): the space or cavity within a tubular structure; such as the lumen in a blood vessel.

lymph (*lymph-,* a clear fluid): a colorless fluid similar to blood plasma containing white blood cells; an important component of the immune and lymphatic system.

lymphocyte (*lymph-,* a clear fluid, + *-cyte,* cell): a type of white blood cell with an agranular cytoplasm that is responsible for immune responses. See agranulocyte (nongranular leukocyte) as well as T cell and B cell.

lysis (*lys-,* to dissolve): the disintegration of a cell due to the rupture of its cell membrane. See autolysis or hydrolysis.

lysosome (*lys-,* to dissolve, + *-som-,* body): an intracellular membrane-bound organelle containing digestive enzymes. These powerful hydrolytic enzymes act when the lysosome ruptures or fuses with a cytoplasmic vesicle. See phagocytosis.

lysozyme (*lys-,* to dissolve, + *-zym-,* enzyme): an enzyme that destroys bacterial cell walls; found in tears, saliva, and sweat.

M

M phase: the phase in the cell cycle that includes mitosis and cytokinesis. See cell cycle and mitosis.

macrandrous (*macr-,* large, + *-andr-,* male, + *-ous,* pertaining to): pertaining to plants or plant-like organisms having male plants or male reproductive structures larger than the female counterparts.

macroevolution (*macr-*, large, + *-evolut-*, an unrolling, + *-ion*, process of): large-scale evolutionary change leading to the formation of new species; also referred to as adaptive radiation, and speciation.

macromolecule (*macr-*, large, + *-mole-*, mass, + *-ule*, little): an extremely large molecule (often a polymer) composed of a repeating combination of smaller molecules. The four specific types of macromolecules include proteins, lipids, nucleic acids, and carbohydrates.

macronutrient (*macr-*, large, + *-nutri-*, to nourish, + *-ent*, having the quality of): any element required in relatively large amounts for normal growth and development including C, H, O, N, P, K, Mg, Ca, S. May also be referred to as an essential element. Opposite of micronutrient.

macrophage (*macr-*, large, +, *-phage*, to eat): a phagocytic white blood cell (enlarged monocyte) that engulfs and digests foreign material and cellular debris. See monocyte.

macrophyll—see megaphyll

malignant (tumor) (*malig-*, bad, + *-ant*, that which, + *tum-*, to swell, + *-or*, state of): a cancerous growth characterized by extremely rapid cell division, abnormal chromosome numbers, and the ability of spreading through the body (metastasis). Opposite of benign.

mantle (*mentel*, a garment): a fleshy fold of tissue that covers the internal organs of mollusks; usually responsible for secreting the shell, as in clams.

marsupial (*marsup-*, pouch, little bag, + *-ial*, pertaining to): a pouched mammal, including kangaroos, opossums, and koalas. The pouch (marsupium) is used to carry the young until their development is complete.

mass extinction: a natural process in which large numbers of species become extinct in a relatively short period of geological time; usually associated with a large-scale environmental disaster. Opposite of background extinction.

mass flow hypothesis: the proposed mechanism for the large-scale movement of phloem sap that states the movement is due to a difference in the hydrostatic pressure at each end of a seive tube. Also referred to as bulk flow hypothesis and pressure flow hypothesis.

matrix (*matri-*, mother): (1) the intercellular semi-viscous material in which cells and tissues are embedded; nonliving and often containing a dense network of fibers. (2) the viscous liquid in the interior of mitochondria in which the cristae are embedded; contains numerous enzymes and coenzymes involved with the Krebs cycle.

mechanical isolation: a prezygotic isolating mechanism that prevents different species from mating because of structural or biochemical differences.

mechanoreceptor (*mechan-*, contrivance, + *-recept-*, to receive, + *-or*, state of): a sensory cell or organ capable of detecting and responding to a physical stimuli; such as the sensory cells associated with touch, pressure, balance, and hearing.

medial (*medi-*, middle, + *-al*, pertaining to): pertaining to the middle of an organ or organism.

mediated transport: a transport mechanism that moves substances across cell membranes with the aid of a carrier molecule; often the carrier is an enzymatic protein called permease.

medulla (*medull-,* marrow): the innermost part of an organ or structure; in contrast to and often covered by the outerpart or cortex; as in the kidney. In plants, the term pith is synonymous with medulla. See cortex.

medulla oblongata (*medull-,* marrow, + *oblong,* + *-ate,* characterized by having): the lowest portion of the brain that is responsible for regulating internal organs; the region where the brain and the spinal cord fuse; also referred to as the brain stem.

megagametophyte (*mega-,* large, + *-gamet-,* union, + *-phyt-,* plant): a female gametophyte arising from a megaspore; also referred to as the embryo sac in angiosperms. Opposite of microgametophyte.

megapascal (named after the French mathematician, B. Pascal): the pascal is the standard unit used to measure pressure. The megapascal is usually used in biology. One megapascal is equal to 10 bars; the bar unit is now obsolete.

megaphyll (*mega-,* large, + *-phyll,* leaf): a large, broad, and flat leaf with numerous branching veins; its leaf trace typically leaves a leaf gap in the stem stele. Also known as macrophyll. Opposite of microphyll.

megasporangium (*mega-,* large, + *-spor-,* spore, + *-angi-,* container, + *-um,* structure): a sporangium producing one to four megaspores. Opposite of microsporangium.

megaspore (*mega-,* large, + *-spor-,* spore): a haploid spore that develops into a female gametophyte; often larger than male microspores. Opposite of microspore.

megasporocyte (*mega-,* large, + *-spor-,* spore, + *-cyte,* cell): a diploid cell that undergoes meiosis to produce four haploid megaspores; also known as a megaspore mother cell. Opposite of microsporocyte.

megasporophyll (*mega-,* large, + *-spor-,* spore, + *-phyll,* leaf): a leaf-like structure bearing megasporangia. Opposite of microsporophyll.

meiosis (*mei-,* reduction, + *-sis,* the process of): a type of nuclear division occurring only in organisms that reproduce sexually, involving two successive nuclear divisions, resulting in the number of chromosomes being reduced to one half the normal number; one single diploid (2n) cell forms four haploid (n) cells (gametes or spores). Meiosis may also be referred to as reduction division.

meiospores (*mei-,* reduction, + spores): haploid spores produced by meiotic cell division. Virtually all spores produced by plants are meiospores.

melanin (*melan-,* black, + *-in,* chemical substance): a dark brown or black pigment common in the outer coverings of animals and responsible for the coloration of skin, hair, and the iris of the eyes.

membrane—see cell membrane

membrane permeability—see semipermeable

memory B cell: a long-lived cell formed by dividing B cells that provides active immunity (secondary immune response) against a specific antibody.

Mendelian genetics (Mendelism) (named after the Austrian Monk, Gregor Mendel): a branch of genetics dealing with inherited traits that are determined by single genes and that have effects easily recognizable. See segregation (Mendel's first law) and independent assortment (Mendel's second law).

Mendel's first law—see segregation

Mendel's second law—see independent assortment

meninges, *sing.* meninx (*mening-*, membrane): the three membranes that cover and protect the brain and spinal cord, including the dura mater, arachnoid, and pia mater.

menopause (*men-*, month, + *-pau-*, to cease): the termination of the menstrual cycle in middle-aged women (usually 45–55 years of age) involving the cessation of egg and hormone production.

menstruation (*men-*, month, + *-stru-*, to flow, + *-ation*, the process of): the monthly discharge of blood and tissue from the uterus through the vagina at the beginning of each menstrual cycle. Also referred to as period or menses.

meristem (*merist-*, divisible): a group or layer of undifferentiated plant cells that activily divide by mitosis to produce new cells. See apical meristem and lateral meristem.

mesenchyme (*mes-*, middle, + *-en-*, in, + *-chym-*, juice, infusion): embryonic connective tissue usually embedded in a jelly-like matrix and that is derived from mesoderm; common in the embryos of vertebrates and in some invertebrate adults.

mesentery (*mes-*, middle, + *-enter-*, gut, + *-y*, state of): a thin sheet of connective tissue that surrounds, supports, and suspends internal organs within the coelom; a part of the peritoneum. See peritoneum.

mesocarp (*mes-*, middle, + *-carp*, fruit): the middle layer of the pericarp (fruit wall) of angiosperms; lies between the endocarp and exocarp.

mesoderm (*mes-*, middle, + *-derm*, skin): one of the three embryonic germ layers that gives rise to all connective tissues, muscle, blood, kidneys, and several other structures; lies between the ectoderm and endoderm.

mesophyll (*mes-*, middle, + *-phyll*, leaf): photosynthetic tissue located between the upper and lower epidermal cell layers of leaves; usually consisting of a lower spongy layer and an upper palisade layer of cells.

mesophyte (*mes-*, middle, + *-phyt-*, plant): a plant that grows in a moderately moist environment and avoids extremes of drought and moisture.

mesosome (*meso-*, middle, + *-som-*, body): an inward folding of the cell membrane in bacteria that functions in separating the duplicated strands of DNA during cell division. It also functions during regular cellular metabolism.

messenger RNA (mRNA): a type of ribonucleic acid (RNA) that has been transcribed from DNA and serves to carry genetic information from the nucleus to the ribosomes in the cytoplasm.

metabolism (*metabol-*, change, + *-ism*, the process of): the sum total of all chemical reactions occurring in an organism, both catabolic and anabolic.

metameres—see somites

metamerism (*meta-*, after, + *-mer-*, part, + *-ism*, the process of)—**see segmentation**

metamorphosis (*meta-*, change, + *-morph-*, form, + *-sis,* process of): the abrupt change from a larval stage to a mature adult; such as the change from a maggot to a fly.

metaphase (*meta-*, middle, + *-phase,* a stage): the second (middle) stage of mitosis or meiosis during which the chromosomes lie in the equatorial plane of the cell.

metaphloem (*meta-*, middle, + *-phlo-*, bark): primary phloem developing after the protophloem and before the secondary phloem (if present).

metastasis—see malignant (tumor)

metaxylem (*meta-*, middle, + *-xyl-*, wood): primary xylem developing after the protoxylem and before the secondary xylem (if present).

metencephalon (*meta-*, middle, + *-encephal-*, brain): the anterior portion of the vertebrate hindbrain that gives rise to the cerebellum and pons.

microbe (*micr-*, small, + *-bi-*, life): any microscopic life form; usually in reference to bacteria.

microbodies (*micr-*, small, + body): small membrane-bound vesicles containing enzymes that break down substances no longer needed by the cell.

microevolution (*micr-*, small, + *-evolut-*, an unrolling, + *-ion,* process of): subtle, small-scale changes in gene frequencies caused by mutations and genetic recombination; often evident within a relatively short period of time (less than 100 years).

microfilaments (*micr-*, small, + *-fil-*, thread, + *-ment-*, state of being): solid thread-like actin fibers that make up portions of the cytoskeleton as well as muscle cells; contractile in nature and therefore cause cells to change shape.

microgametophyte (*micr-*, small, + *-gamet-*, union, + *-phyt-*, plant): a male gametophyte arising from a microspore; also referred to as a pollen grain in seed plants. Opposite of megagametophyte.

microglia (*micr-*, small, + *-glia-*, glue): a group of relatively small cells within nervous tissue that do not conduct nerve impulses but rather protect the system by removing foreign materials by phagocytosis and producing myelin. See neuroglia cells.

micrometre (*micr-*, small, + *-meter,* a unit of length): a unit of measure equal to 1/1000 of a millimetre; the unit typically used when studying cells. The term micrometre has replaced the obsolete term micron. Symbolized µm.

micronutrients (*micr-*, small, + *-nutri-*, to nourish, + *-ent,* having the quality of): inorganic mineral elements required in minute amounts for normal growth and development; such as Cl, Fe, Cu, Zn, Bo. Also known as trace elements. Opposite of macronutrients.

microphyll (*micr-*, small, + *-phyll,* leaf): a small, scale-like leaf with one vein; its leaf trace is not associated with a leaf gap in the stem stele. Opposite of megaphyll.

micropyle (*micr-*, small, + *-pyl-*, gate): a small opening in the integument(s) of the ovule through which the pollen tube enters.

microspheres—see protenoid microsphere

microsporangium (*micr-*, small, + *-spor-*, spore, + *-angi-*, container, + *-um*, part, region): a sporangium producing microspores. Opposite of megasporangium.

microsome—see peroxisome

microspore (*micr-*, small, + *-spor-*, spore): a haploid spore that develops into a male gametophyte, often smaller than a megaspore. Opposite of megaspore.

microsporocyte (*micr-*, small, + *-spor-*, spore, + *-cyte*, cell): a diploid cell that undergoes meiosis to produce four haploid microspores; also known as a microspore mother cell. Opposite of megasporocyte.

microsporophyll (*micr-*, small, + *-spor-*, spore, + *-phyll*, leaf): a leaf-like structure bearing microsporangia. Opposite of megasporophyll.

microtubules (*micr-*, small, + *-tub-*, tube, + *-ule*, little): long hollow rods of tubulin protein that make up portions of the cytoskeleton and help to maintain cell shape. They are also found in cilia, flagella, and centrioles and make up the characteristic "9+2" arrangement seen in a cross section of eukaryotic cilia and flagella.

microvillus, *pl.* **microvilli** (*micr-*, small, + *-vill-*, hairy, + *-us*, thing): a minute hair-like projection on the cell membrane of some animal cells that increases the surface area of the cell.

middle lamella (middle, + *lamell-*, thin plate): an intercellular layer of pectin between the walls of adjacent plant cells that acts to cement the cells together.

midrib: the central vein of a leaf that provides support and conducts fluids within the leaf.

mitochondrion, *pl.* **mitochondria** (*mito-*, thread, + *-chondr-*, granule): an elongated or spherical membrane-bound organelle in which cellular respiration takes place; often referred to as the "powerhouse" of the cell since most ATP in a eukaryotic cell is produced and stored within it.

mitosis (*mito-*, thread, + *-sis*, the process of): a type of nuclear division involving the replication and distribution of a complete set of identical chromosomes to each daughter cell; for the purpose of growth and repair; also referred to as karyokinesis.

mitotic spindle—see spindle fibers (apparatus)

molecular weight (*mole-*, mass, + *-ule*, little, + *-ar*, pertaining to, + weight): the total weight of all the atoms in a molecule or compound.

molecule (*mole-*, mass, + *-ule*, little): a chemical unit consisting of two or more atoms held together by ionic or covalent bonds.

molting—see ecdysis

monoclonal antibody (*mono-*, one, + *-clon-*, twig, + *-al*, pertaining to, + *anti-*, against + body): a specific antibody produced by a hybrid clone of B cells and cancer cells, from which large amounts of a specific antibody can be obtained for research purposes. A number of viral diseases as well as some forms of cancer are treated with antibodies produced in this manner.

monocotyledon (*mono-*, one, + *-cotyle-*, cup): a group of flowering plants, the Monocotyledonae, with embryos having one cotyledon. Often abbreviated monocot.

monocyte (*mono-*, one, + *-cyte*, cell): a relatively large phagocytic agranular leukocyte that, in the presence of an antigen, enters the tissue and differentiates into a macrophage.

monoecious (*mono-*, one, + *-eci-*, dwelling, + *-ous*, pertaining to): usually refers to plants having male and female sex organs on the same individual. Also known as bisexual or hermaphroditic in animals. Opposite of dioecious.

monohybrid (*mono-*, one, + *-hybrid*, mixed offspring): the offspring of parents who differ in one distinct trait.

monohybrid cross (*mono-*, one, + *-hybrid*, mixed offspring, + cross): a genetic cross between two parents who differ in one distinct trait.

monomer (*mono-*, one, + *-mer-*, a part of): a simple, relatively small molecule that can be bonded to other similar monomers to form a polymer.

monosaccharide (*mono-*, one, + *-sacchar-*, sugar, + *-ide*, denoting a chemical compound): a simple carbohydrate (sugar) made up of one molecule that cannot be broken down by hydrolysis into smaller sugar molecules; such as glucose, ribose, and fructose.

morph (*morph-*, form): an organism that exists in two or more distinct forms. The human liver fluke (*Chlonorchis sinensis*), for example, exhibits three morphs which parasitize three different hosts.

morphogenesis (*morph-*, form, + *-gen-*, origin, + *-sis*, process of): the process of developing form and structure.

morphology (*morph-*, form, + *-logy*, study of): the study of structure, both internal and external at varying levels, including cells, tissues, organs, and organ systems.

morula (*morul-*, a mulberry): one of the earliest embryonic stages of development in animals consisting of a solid ball of cells; a stage prior to the blastula stage.

motor neuron: a nerve cell that carries impulses from the CNS to a gland or muscle.

mRNA—see messenger RNA

multiple alleles: a hereditary mechanism involving a single gene with three or more alternative alleles that control the same trait; such as the ABO blood type in humans.

multiple fruit: one fruit formed from the mature ovaries of more than one flower; such as a pineapple.

mutagen (*muta-*, change, + *-gen*, origin): any agent (chemical or physical) that is capable of altering DNA and thus producing a mutation; such as radiation.

mutation (*muta-*, change, + *-tion*, the process of): an inheritable chemical change in the DNA sequence of a gene.

mutualism (*mutu-*, reciprocal, + *-al-*, pertaining to, + *-ism*, the process of): a relationship between two different species that benefits both; a type of symbiosis. See symbiosis.

mycelium (*myc-*, fungus, + *-ium*, region): the collective term for all the hyphae of a fungus; responsible for breaking down and absorbing nutrients. The vegetative body of a fungus.

mycobiont (*myc-*, fungus, + *-bi-*, life, + *-ont,* being): the fungal partner of a lichen. Opposite of phycobiont. See lichen.

mycology (*myc-*, fungus, + *-logy,* study of): the study of fungi.

mycorrhiza (*myc-*, fungus, + *-rrhiz-*, root): a symbiotic association of a fungus with the roots or rhizomes of higher plants.

mycosis (*myc-*, fungus, + *-osis,* diseased): any disease caused by a fungus; such as "athlete's foot" or "ringworm."

myelencephalon (*myel-*, marrow, + *-encephal-*, brain): the posterior portion of the vertebrate hindbrain containing the medulla oblongata; in an embryo it makes up the entire hindbrain.

myelin (*myel-*, marrow, + in, chemical substance): a fatty material produced by Schwann cells and oligodendrocytes that forms a sheath (myelin sheath) around neuron fibers giving them a white appearance. See white matter.

myofibrils (*myo-*, muscle, + *-fibr-*, fiber): thread-like organelles composed of contractile proteins (actin and myosin) found in the cytoplasm of striated and cardiac muscle cells; responsible for muscle contraction.

myofilament (*myo-*, muscle, + *-fil-*, thread, + *-ment,* state of being): one of the thread-like filaments of a myofibril; the structural protein unit of a muscle cell.

myoglobin (*myo-*, muscle, + *-glob-*, sphere, + *-in,* chemical substance): a globular protein pigment found in muscle that transports and stores oxygen.

myometrium (*myo-*, muscle, + *-metr-*, mother, uterus, + *-ium,* region): the layer of smooth muscle in the wall of the uterus. Opposite of endometrium.

myosin (*myo-*, muscle, + *-in,* chemical substance): a globular protein that, along with actin, is responsible for muscle contraction; the major component of thick filaments in myofibrils. See actin.

N

NAA (α-naphthaleneacetic acid): a synthetic auxin. See auxins.

NAD (Nicotinamide Adenine Dinucleotide): an important coenzyme that is oxidized or reduced to serve as an electron acceptor or donor in the Krebs cycle and other metabolic oxidation-reduction reactions.

NADP (Nicotinamide Adenine Nucleotide phosphate): an important coenzyme that accepts (reduced) or donates (oxidized) hydrogen atoms in various metabolic reactions; a type of hydrogenase enzyme.

nastic response (*nast-*, pressed close, + *-ic*, pertaining to, + response): movement in plants caused by sudden changes in the turgor pressure of petiole cells; such as the rapid movement seen in some insectivorous plants ("flytraps").

natural selection: the evolutionary process by which organisms better adapted to a particular environment are favored to reproduce to a greater degree than less adapted organisms and thereby pass on their genes to the next generation; also referred to as differential, or nonrandom mating. See Darwinism.

necrosis (*necr-*, death, + *-sis,* process of): the death of tissue in a localized region due to injury, disease, or a mineral deficiency; usually in reference to plants.

nectary (*nectar-*, drink of the gods, + *-y*, state of): glandular cells usually in flowers that secrete a sugar-rich solution (nectar) used to attract pollinators.

nematocyst (*nemat-*, thread, + *-cyst*, sac): a long, thread-like stinger characteristic of cnidarians (''jelly-fish'' and other related organisms) used for defense, anchorage, and capturing prey; contained within a special cell called a cnidocyte.

neo-Darwinism (*neo-*, new, young, + Darwinism; named after the English naturalist, C. Darwin): the relatively modern theory of evolution based on recent discoveries in population genetics, Mendelian genetics, and biochemistry, in addition to Darwin's original idea of natural selection. Neo-Darwinism is also referred to as the modern synthetic theory of evolution, since it is a theory based on the ''synthesis'' of several ideas.

neoteny (*neo-*, new, young, + *-ten-*, something stretched, + *-y*, state of): the retention of juvenile characteristics in adult organisms, or sexual maturity occurring in the larval stage.

nephridium (*nephr-*, kidney, + *-id-*, pertaining to, + *-ium*, little): a primitive excretory organ consisting of a ciliated funnel and tubule which drain the coelom to the exterior; common to many invertebrates.

nephron (*nephr-*, kidney): the functional and structural unit of the vertebrate kidney, made up of a glomerulus enclosed within Bowman's capsule and a complex network of tubules including the loop of Henle.

nerve cell—see neuron

net productivity: the total amount of chemical energy (in kilojoules) available to primary consumers in an ecosystem; a value equal to gross production minus respiration and metabolism by primary producers.

net venation—see reticulate-net venation

neuroglia cells (*neuro-*, nerve, + *-glia-*, glue, + cells): specialized cells within the Central Nervous System that do not conduct nerve impulses but rather support and nourish neurons. The three types of neuroglia cells include astrocytes, oligodendrocytes, and microglia.

neurolemma (*neuro-*, nerve, + *-lemma*, sheath): a thin, nucleated covering on nerve cells that enhances the conduction of nerve impulses; also referred to as the sheath of Schwann.

neuron (*neuro-*, nerve): a nerve cell consisting of a cell body, axon(s), and dendrite(s); the fundamental unit of the nervous system.

neurosecretory cell (*neuro-*, nerve, + *-secret-*, to separate, + *-ory*, place for, + cell): any cell within the nervous system that produces and secretes a hormone; such as the cells of the hypothalamus.

neurotransmitter (*neuro-*, nerve, + transmitter): a chemical compound produced at the ends of neurons used for transmitting nerve impulses across a synapse (gap between adjacent nerve cells); such as acetylcholine.

neutron (*neutr-*, neither one nor the other, + *-on*, a particle): a primary subatomic particle with no electrical charge; found with one or more protons in the nucleus of an atom.

neutrophil (*neutr-*, neither one nor the other, + *phil*, to love): a primary phagocyte. See granulocyte.

niche (*nich,* to nest)—**see ecological niche**
nidamental gland (*nid-,* nest, + *-ment-,* condition of, + *-al,* pertaining to): a gland that secretes protective and nutritive coverings for eggs.
nitrification (*nitri-,* soda, + *-fic-,* to make, + *-ation,* the process of): the oxidation of ammonia (NH_3) or ammonium (NH_4^+) to nitrates (NO_3^-) or nitrites (NO_2^-) by certain types of nitrifying bacteria.
nitrogen fixation: the unique metabolic ability of some bacteria and blue-green algae to convert free atmospheric nitrogen (N_2) into usable nitrogenous compounds such as ammonia (NH_3); often followed by nitrification.
node (*nod-,* knob): the part of a stem where one or more leaves are attached. Opposite of internode.
nodules (*nod-,* knob, + *-ule,* small): tumor-like swellings on the roots of legumes in response to the presence of nitrogen-fixing bacteria.
nonbiodegradable (*non-,* not, + *-bi-,* life, + *-degrad-,* to break down, + *-able,* able to): any substance that can not be decomposed by living organisms; such as plastic. Opposite of biodegradable.
noncompetitive inhibitor: a chemical that, when bound to an enzyme (other than at the active site), reduces or destroys the enzyme's ability to function by changing the active site so the substrate can no longer bind.
nondisjunction (*non-,* not, + *-dis-,* apart, + *-junct-,* to join, + *-ion,* the process of): the abnormal separation (or complete failure of separation) of homologous chromosomes or sister chromatids during meiosis.
norepinephrine (*nor-,* a normal or parent compound, + *-epi-,* upon, + *-nephr-,* kidney, + *-in,* chemical substance): a hormone produced by the adrenal gland that can function as a stimulatory hormone (increases heart rate, raises blood pressure) and as a neurotransmitter. Also referred to as noradrenalin.
notochord (*noto-,* back, + *-chord,* cord): a flexible, longitudinal supporting rod located just ventral to the nerve cord in all chordates; replaced by the vertebral column in vertebrates.
nucellus (*nucell-,* a small nut, + *-us,* thing): the inner part of the ovule in seed plants in which the embryo sac develops; often considered equivalent to the megasporangium.
nuclear envelope (*nucle-,* nucleus, kernel, + *-ar,* pertaining to, + *en-,* in, + *-velop-,* to wrap): the double membrane that surrounds the nucleus and separates it from the cytoplasm; often continuous with the endoplasmic reticulum. Also referred to as the nuclear membrane.
nucleic acid (*nucle-,* nucleus, kernel, + *-ic,* pertaining to, + acid): a macromolecule made up of nucleotide subunits; the two types are deoxyribonucleic acid (DNA) and ribonucleic acid (RNA). See deoxyribonucleic acid and ribonucleic acid.
nucleoid (n*ucle-,* nucleus, kernel, + *-oid,* resembling): an unbounded region in prokaryotic cells consisting of a concentrated mass of DNA.
nucleolus (*nucle-,* nucleus, kernel, + *-ole-,* little, + *-us,* thing): a relatively small specialized organelle within the nucleus; mainly composed of RNA and the site of ribosome and rRNA manufacture.

nucleosome (*nucle-*, nucleus, kernel, + *-som-*, body): a bead-like unit consisting of a segment of DNA coiled around eight histone protein molecules; the typical unit eukaryotic DNA is packaged in.

nucleotide (*nucle-*, nucleus, kernel, + *-ide*, pertaining to): an organic molecule consisting of a 5-carbon sugar (either ribose or deoxyribose), a nitrogenous base, and a phosphate group; a "building block" (monomer) of nucleic acids.

nucleus (*nucle-*, nucleus, kernel, + *-us*, thing): (1) a membrane-bound organelle containing DNA; often referred to as the control center of the cell. (2) the central region of an atom containing protons and neutrons.

nut: a hard, dry, nonsplitting, one-seeded fruit; such as a hazelnut.

O

obligate aerobe (*oblig-*, obliged, + *-ate*, characterized by having, + *aer-*, air, + *-bi-*, life): an organism that requires molecular oxygen for cellular respiration.

obligate anaerobe (*oblig-*, obliged, + *-ate*, characterized by having, + *an-*, without, + *-aer-*, air, + *-bi-*, life): a bacterium that does not require molecular oxygen for cellular respiration and is poisoned by it; such bacteria will usually generate ATP through the process of fermentation.

occlusion (*occlu-*, shut up, + *-ion*, process of): (1) a process blocking an opening to a duct or vessel; such as a stroke that is blockage of blood flow to the brain by an occlusion in cerebral blood vessels. (2) the meeting of the teeth when the jaw is closed.

ocellus (*ocel-*, small eye, + *-us*, thing): a simple light receptor or light spot common to invertebrates; such as *Daphnia*.

Okazaki fragments (named after the Japanese geneticist, R. Okazaki): short discontinuous segments of the lagging strand of DNA synthesized in the 5′ to 3′ direction.

olfactory (*olfac-*, smell, + *-ory*, place for): pertaining to the sense of smell, such as the olfactory nerve.

oligodendrocyte (*olig-*, few, + *-dendr-*, tree-like, + *-cyte*, cell): a relatively rare neuroglial cell that is responsible for producing myelin in the Central Nervous System. See neuroglia cells.

ommatidium, *pl.* ommatidia (*omma-*, eye, + *-idium*, little): an individual optic unit found in the compound eyes of arthropods and mollusks. See compound eye.

omnivore (*omni-*, all, + *-vor-*, to eat): an animal that eats both plants and animals; such as humans.

oncogene (*onc-*, a tumor, + *-gen-*, origin): a gene that may cause cancer when activated, by coding for regulatory proteins involved in the control of cell growth and division; often part of an organisms normal genetic makeup (genome) or part of viral genomes.

ontogeny (*ont-*, individual, + *-gen-*, origin, + *-y*, process of): the developmental history of an individual organism. Opposite of phylogeny.

oocyte (*oo-*, egg, + *-cyte*, cell): an animal cell that produces an egg (ovum) cell by meiosis.

oogamy (*oo-*, egg, + *-gam-*, union, + *-y*, process of): a type of sexual reproduction in which the egg is large and sessile, and the sperm are small and motile. Opposite of isogamy.

oogonium (*oo-*, egg, + *-gon-*, reproduction, + *-ium*, part, region): a unicellular female gametangium containing one egg; common in certain algae and fungi.

operator (*oper-*, work, + *-or*, result of the act of): a regulatory segment of DNA that controls the transcription of a gene into mRNA by interacting with repressor proteins.

operculum (*opercul-*, a lid, + *-um*, structure): any lid or flap-like covering; such as the top of the spore capsule in mosses, or the gill cover of fishes.

operon (*oper-*, work, + *-on*, particle): a segment of DNA in prokaryotic cells consisting of a cluster of structural genes that are controlled by one operator and one repressor.

optic (*opt-*, eye, + *-ic*, pertaining to): pertaining to the eye; such as the optic nerve.

oral (*or-*, mouth, + *-al*, pertaining to): pertaining to the mouth or oral region of an animal.

organelle (*organ-*, tool, instrument, + *-elle*, little): a specialized membrane-bound structure within cells that performs a specific function; such as the nucleus, endoplasmic reticulum, or golgi complex.

organic compound (*organ-*, tool, instrument, + *-ic*, pertaining to, + compound): any compound containing carbon; often the atoms involved are covalently bonded.

organogenesis (*organ-*, tool, instrument, + *-gen-*, origin, + *-sis*, process of): an early stage of embryonic development during which organ development occurs.

origin (of a muscle): the end of a muscle that attaches to a relatively immovable bone. Opposite of insertion.

osmoregulation (*osmo-*, pushing, + *-regul-*, rule, + *-tion*, the process of): the process of controlling the internal water balance in living organisms by maintaining a proper internal solute/solvent ratio.

osmosis (*osmo-*, pushing, + *-sis*, the process of): the movement of water across a selectively permeable membrane from a region of high concentration to a region of low concentration; a special type of diffusion reserved for the movement of water since water is the principal biological solvent.

osmotic potential (*osmo-*, pushing, + *-ic*, pertaining to, + potential): a measure of the tendency for water to move across a selectively permeable membrane to regions of lower solute concentration; the higher the solute concentration of a solution, the greater the osmotic potential.

osteocyte (*oste-*, bone, + *-cyte*, cell): a mature bone cell located within a cavity (lucuna); responsible for maintaining and repairing bone tissue.

ostiole (*osti-*, mouth-like opening, + *-ole*, little): a general term for any opening or pore; such as the opening of the conceptacle of the brown alga *Fucus*, or the intake pore of sponges; synonymous with ostium.

ovary (*ov-*, an egg, + *-ary*, place for): (1) a female sex organ that produces eggs (female gametes) and sex hormones. (2) the lower swollen portion of the pistil that contains one or more ovules and that eventually develops into a fruit.

oviparous (*ov-*, an egg, + *-par-*, give birth to, + *-ous,* pertaining to): pertaining to an animal bearing young animals that hatch from eggs laid outside the mother's body; birds are oviparous.

ovoviviparous (*ov-*, an egg, + *-viv-*, alive, + *-par-*, give birth to, + *-ous,* pertaining to): pertaining to an animal bearing young that hatch from eggs that are retained within the mother's body; a number of sharks are ovoviviparous.

ovule (*ovul-*, a little egg): the megasporangium in seed plants consisting of the megagametophyte, nucellus, and integuments and that eventually develops into a seed.

ovum (*ov-*, an egg, + *-um,* structure): an egg or mature female gamete; usually used in a zoological context.

oxidation (*oxida-*, to oxidize, + *-tion,* the process of): the loss of one or more electrons by an atom or molecule (in reference to inorganic compounds) and the loss of one or more hydrogen ions (in reference to organic compounds). Opposite of reduction. Oxidation and reduction occur simultaneously with the electron lost by one molecule immediately being taken up by another molecule.

oxidation-reduction reaction: a common reaction in which one molecule loses electrons (oxidation) while another molecule simultaneously gains electrons (reduction). Also referred to as redox reaction.

oxidative phosphorylation—see electron transport chain

ozone (*oze-*, to smell): O_3; a gas that forms a layer in the upper atmosphere and shields the earth from most ultraviolet radiation. Ozone has a pungent odor similar to chlorine.

P

P-site: the location on a ribosome where tRNA, carrying a growing polypeptide chain, attaches.

P_1 (parental) generation: the two parents that produce the first generation (F_1) of offspring in a genetic cross.

pachytene (*pachy-*, thick, + *-ten-*, to hold): the third stage of prophase I in meiosis during which chromatin begins to condense into chromosomes.

paleobotany (*pale-*, ancient, + *-botan-*, grass): the study of fossilized plants.

paleontology (*pale-*, ancient, + *-ont-*, individual, + *-logy,* study of): the study of fossils.

palisade layer: vertically arranged photosynthetic cells located below the upper epidermis in leaves; a portion of the mesophyll layer. See mesophyll.

panicle (*panicl-*, a small tuft): a type of branched inflorescence in which loose clusters of flowers occur at the tips of each branch.

papilla, *pl.* **papillae** (*papill-*, nipple): any small cone or nipple-like projection, such as the papilla at the base of a hair follicle.

parallel evolution (*para-*, along side, + *evolut-*, an unrolling, + *-ion*, process of): the evolution of similar characteristics in distantly related groups as a result of being exposed to similar environmental conditions; such as the sabertooth-like cats of North America and the sabertooth marsupials of South America.

parallel venation (*para-*, along side, + *ven-*, vein, + *-ation*, the process of): an arrangement of leaf veins typical of monocots in which the major veins are parallel to the longitudinal axis of the leaf.

paraphysis, *pl.* paraphyses (*para-*, along side, + *-phy-*, to grow, + *-sis*, the process of): a sterile hair-like projection growing between and separating reproductive structures in the sporangia or gametangia of primitive plants; such as algae and mosses.

parasite (*para-*, along side, + *-sit-*, food): an organism that feeds off of and derives its energy from a living host; such as certain fungi and worms.

parasitism (*para-*, along side, + *-sit-*, food, + *-ism*, the process of): an intimate association between two or more different organisms in which one benefits and the other may be harmed, or at least weakened; a type of symbiosis. See symbiosis.

parasympathetic nervous system (*para-*, alongside, + *-sympath-*, of like feelings, + *-ic*, pertaining to, + nervous system): pertaining to the portion of the Autonomic Nervous System that is concerned with conserving and restoring energy following emergencies, such as slowing heart rate. Opposite of the sympathetic nervous system.

parenchyma (*para-*, along side, + *-en-*, in, + *-chym-*, infusion): relatively unspecialized plant tissue composed of thin-walled cells with large vacuoles primarily used for storage; the most common type of plant tissue.

parietal (*pariet-*, wall, + *-al*, pertaining to): pertaining to the wall of a cavity.

parthenocarpy (*parthen-*, a virgin, + *-carp-*, fruit, + *-y*, process of): the development of fruit without fertilization; often the result of the exogenous application of plant hormones.

parthenogenesis (*parthen-*, a virgin, + *-gen-*, origin, + *-sis*, process of): the development of an organism from an unfertilized egg; common in bees, wasps, and aphids.

passive transport: a transport mechanism not requiring cellular energy (ATP) that moves substances across a cell membrane down a concentration gradient (from a high concentration to a low concentration). See diffusion and osmosis.

pathogen (*path-*, disease, + *-gen*, origin): any organism capable of causing disease; usually in reference to bacteria, but viruses also apply.

pectin (*pect-*, congealed, + *-in*, chemical substance): an organic polymer composed of calcium pectate and pectic acid that serves as a cementing agent within the middle lamella of plant cell walls.

pedicel (*ped-*, foot): the stalk or stem of an individual reproductive structure; as on a flower or gametangium.

peduncle (*peduncl*, a little foot): (1) any stem or stalk-like structure; such as the stem of a strobilis or inflorescence. (2) a band of tissue connecting parts of the brain.

pelagic (*pelag-*, the sea, + *-ic*, pertaining to): pertaining to living in the open ocean.

pellicle (*pell-*, skin, + *-cle*, little): an ornamented protein layer just inside the cell membrane of certain protozoans.

pentose (*pent-*, five, + *-ose*, sugar): a 5-carbon sugar such as ribose and deoxyribose; a major component of nucleic acids.

peptidase (*pept-*, to digest, + *-ase*, enzyme): a hydrolytic enzyme that breaks down dipeptides and polypeptides (proteins) into amino acids.

peptide (*pept-*, to digest, + *-ide*, denoting a chemical compound): a long-chain molecule composed of amino acids; a dipeptide consists of two amino acids and a polypeptide consists of many amino acids; also referred to as a protein.

peptide bond (*pept-*, to digest, + *-ide*, denoting a chemical compound, + bond): a type of covalent bond resulting from a dehydration reaction that joins two amino acids together to form a dipeptide; three or more amino acids linked in this way form a polypeptide chain (protein).

peptidoglycan (a contraction of peptide and glycogen): a complex polymer composed of carbohydrate units bonded to proteins; found only in the cell walls of Gram positive bacteria.

perennial (*per-*, through, + *-annu-*, year, + *-al*, pertaining to): a plant that lives for many years and that may or may not reproduce each year.

perfect flower: a flower that has both stamens and a pistil; also referred to as bisexual or monoecious. Opposite of imperfect flower.

perianth (*peri-*, around, + *-anth*, flower): a collective term for all the sepals and petals of a flower.

pericarp (*peri-*, around, + *-carp*, fruit): the fruit wall that develops from the mature ovary wall.

periclinal (*peri-*, around, + *-clin-*, to lean, + *-al*, pertaining to): a plane of cell division parallel to the surface of an apical meristem or other surface. Opposite of anticlinal.

pericycle (*peri-*, around, + *-cycle*, ring): a layer of meristematic cells internal to the endodermis and surrounding the xylem and phloem of roots; the site of lateral root formation.

periderm (*peri-*, around, + *-derm*, skin): a collective term for a layer of protective tissues that replace the epidermis in woody plants; composed of cork, cork cambium, and phelloderm.

Peripheral Nervous System (PNS) (*peri-*, around, + *-pher-*, to carry, + *-al*, pertaining to, + nervous system): the entire vertebrate nervous system excluding the brain and spinal cord; composed of two major subdivisions: the Autonomic Nervous System and the Somatic Nervous System.

peripheral proteins (*peri-*, around, + *-pher-*, to carry, + *-al*, pertaining to, + *prote-*, first, + *-in*, chemical substance): protein molecules found on the surface of cell membranes; important in cellular communication and cell recognition.

peristalsis (*peristal-*, compressing around, + *-sis*, process of): rhythmic wave-like contractions of smooth muscle that move food through the digestive tract, or that move eggs through the oviduct in vertebrates.

peristome (*peri-*, around, + *-stom-*, mouth): a fringe of teeth-like structures surrounding the opening of moss spore capsules that aid in spore dispersal.

perithecium (*peri-*, around, + *-thec-*, case, + *-ium*, region): a round or flask-shaped ascocarp with a small opening or ostiole at the top.

peritoneum (*periton-*, stretched over, + *-um*, structure): a membrane of connective tissue that lines the abdominal cavity and covers the visceral organs.

permeable (*perm-*, to pass through, + *-able*, able to): allowing the passage of substances such as ions or molecules; usually in reference to cell membranes. See semipermeable.

peroxisome (*peroxide*, + *-som-*, body): a membrane-bound organelle containing enzymes that produce or destroy hydrogen peroxide (H_2O_2); also found in plant cells and is the site of photorespiration. Also referred to as a microsome.

petal (*petal-*, leaf): a sterile, innermost portion of the perianth of a flower; often conspicuously colored in order to attract pollinators.

petiole (*petiol-*, small leaf): the stem-like stalk of a leaf.

pH (scale) (**p**otential of **H**ydrogen): a measure of the concentration of protons (hydrogen ions) in a solution. The pH scale runs from 0 to 14 with basic (alkaline) solutions having a pH greater than 7, and acidic solutions having a pH less than 7; pH 7 is neutral. See acid, and base.

phage—see bacteriophage

phagocyte (*phag-*, to eat, + *-cyte*, cell): usually in reference to a type of white blood cell that engulfs and digests foreign material (bacteria) and cellular debris. See macrophage.

phagocytosis (*phag-*, to eat, + *-cyt-*, cell, + *-sis*, process of): the process of engulfing relatively large particles of foreign material and cellular debris by cells. See endocytosis. Opposite of exocytosis.

pharynx, ***pl.*** **pharynges** (*pharyn-*, throat): a region where the digestive and respiratory tracts of vertebrates meet; located between the oral cavity and the esophagus.

phellem—see cork

phelloderm (*phell-*, cork, + *-derm*, skin): a protective layer of cells produced inwardly by the cork cambium and opposite to the cork layer.

phellogen—see cork cambium

phenotype (*phen-*, visible, + *-typ-*, form): the genetically expressed characteristics of an organism that are determined by the organism's genotype and the environment to which the organism is exposed. See genotype.

pheromone (*pher-*, to carry, + *hormone*, to excite): an extremely potent chemical secreted by one organism that influences the behavior and physiology of another organism of the same species; commonly used by insects to attract mates.

phloem (*phlo-*, bark): living, food-conducting vascular tissue composed of sieve tube cells, companion cells, parenchyma, and fibers. Phloem tissue is usually located outside of and surrounding xylem tissue. Together, phloem and xylem make up the stele.

phospholipid (a contraction of phosphorous and lipid): a type of lipid in which the phosphate "head" (hydrophilic) is bonded to a lipid "tail" (hydrophobic). Such a molecule is amphipathic; a fundamental component of cell membranes.

phosphorylation (*phosphor-*, bringing light, + *-yl-*, chemical radical, + *-ation*, the process of): the process of adding a phosphorous atom or phosphate group to a compound; such as the formation of ATP by adding a phosphorous atom to ADP.

photoautotrophic (*photo-*, light, + *-aut-*, self, + *-troph-*, to feed, + *-ic*, pertaining to): the ability to synthesize organic compounds from inorganic compounds and light energy.

photolysis (of water) (*photo-*, light, + *-lys-*, to dissolve): the splitting of water molecules into oxygen, electrons, and protons using light energy; an essential part of the light-dependent phase of photosynthesis.

photon (*photo-*, light): an elementary unit of light energy that cannot be further subdivided. Also known as a quantum.

photoperiodism (*photo-*, light, + *-period-*, a going around, + *-ism*, the process of): plant growth and development in response to the length of day and/or night; usually in reference to reproduction.

photophosphorylation (*photo-*, light, + *-phosphor-*, bringing light, + *-yl-*, chemical radical, + *-ation*, the process of): the formation of ATP using light energy as opposed to using chemical energy as in oxidative phosphorylation; occurs in chloroplasts.

photorespiration (*photo-*, light, + *-respir-*, to breathe, + *-ation*, the process of): a relatively inefficient metabolic reaction associated with the dark reactions of photosynthesis in which no carbohydrates are produced; occurs during hot, dry, and sunny conditions when stomata are closed.

photosynthesis (*photo-*, light, + *-syn-*, to put together, + *-the-*, in place, + *-sis*, process of): the process in which light energy and chlorophyll are used to manufacture carbohydrates from carbon dioxide and water.

photosystem (I and II) (*photo-*, light, + *-system*, a composite whole): two molecular systems used for the conversion of light energy into chemical energy. Photosystem I (PS I) absorbs light in the 700 nm range. Photosystem II (PS II) absorbs light in the 680 nm range.

phototropism (*photo-*, light, + *-trop-*, to turn, + *-ism*, the process of): a direction of growth in response to a unidirectional light source. Shoots are positively phototropic whereas roots are negatively phototropic.

phycobilins (*phyc-*, algae, + *-bil-*, bile, + *-in*, chemical substance): a group of red and blue photosynthetic pigments occurring in blue-green algae and red algae; chemically similar to bile. See phycocyanin and phycoerythrin.

phycobiont (*phyc-*, algae, + *-bi-*, life, + *-ont*, individual): the algal partner of a lichen. Opposite of mycobiont. See lichen.

phycocyanin (*phyc-*, algae, + *-cyan-*, blue + *-in*, chemical substance): a blue phycobilin pigment occurring in blue-green algae and red algae.

phycoerythrin (*phyc-*, algae, + *-erythr-*, red + *-in*, chemical substance): a red phycobilin pigment occuring in blue-green algae and red algae.

phycology (*phyc-*, algae, + *-logy,* the study of): the study of algae.
phyletic evolution—see anagenesis
phyllotaxy (*phyll-*, leaf, + *-tax-*, arrangement, + *-y,* condition of): the arrangement of leaves on a stem.
phylogeny (*phyl-*, race, + *-gen-*, origin, + *-y,* process of): the complete evolutionary history of a species or group of organisms; often represented as an "evolutionary tree." Opposite of ontogeny.
physiology (*physi-*, function, + *-logy,* study of): the study of the functions and processes performed by organisms.
phytochrome (*phyt-*, plant, + *-chrom-*, color): a cytoplasmic pigment associated with the absorption of red and far-red wavelengths; involved with numerous timing processes including flowering, seed germination, dormancy, and leaf formation.
phytoplankton (*phyt-*, plant, + *-plank-*, drifting, + *-on,* a particle): free-floating or flagellated microscopic aquatic plants and algae; autotrophic plankton.
pigment (*pig-*, to paint, + *-ment,* result of): a colored organic compound in plants or animals that absorbs light.
pileus (*pile-*, cap, + *-us,* thing): the cap-like top of basidiocarps ("mushrooms").
pili (*pil-*, hair): appendage-like extensions on the surface of certain bacteria used to transfer DNA during conjugation.
pinna (*pinn-*, feather): (1) a subdivision of a compound leaf or fern frond. See leaflet. (2) the outer ear.
pinnate (*pinn-*, feather, + *-ate,* characterized by having): a type of leaf venation with a major vein running down the center of the leaf and branches occurring at uniform angles along the central axis. A morphology similar to a feather.
pinocytosis (*pino-*, drink, + *-cyt-*, cell, + *-sis,* the process of): the process of engulfing and absorbing droplets of water (containing dissolved solutes) by cells. See endocytosis. Opposite of exocytosis.
pistil (*pistil-*, pestle): the female part of a flower consisting of a stigma, style, and ovary. A pistil may contain one or more carpels.
pistillate (*pistil-*, pestle, + *-ate,* characterized by having): referring to a unisexual flower having one or more pistils but no stamens; also known as carpellate.
pit: any depression or opening; usually in reference to the small openings in the cell walls of xylem cells that function in providing a continuum between adjacent xylem cells.
pith: parenchyma tissue in the center of some stems and roots.
placenta (*placent-*, small flat cake): (1) a structure that connects an embryo to the surrounding maternal tissue and through which materials are exchanged between the embryo and mother; tissue that is partially maternal and partially embryonic; also referred to as afterbirth. (2) the region of the ovary wall to which the ovules or seeds are attached; vaguely analagous to the placenta of mammals.
placentation (*placent-*, small flat cake, + *-ation,* the process of): the specific pattern of ovule attachment to the ovary wall.
plankton—see phytoplankton

planula, ***pl.*** **planulae** (*plan-*, wandering, + *-ula*, little): the ciliated, free-swimming larval stage of many cnidarians.

plaque (*plaque*, plate): (1) an area of growth in a bacterial culture cleared by a bacteriophage. (2) a deposit of fatty connective tissue in the inner wall of blood vessels; associated with athersclerosis.

plasma (*plasm-*, formed): the clear liquid portion of blood in which the cells are suspended.

plasma cell: a fully differentiated B cell that actively produces antibodies in response to specific antigens.

plasmalemma—see cell membrane

plasma membrane—see cell membrane

plasmid (*plasm-*, formed, + *-id*, tending to): a circular segment of DNA found only in prokaryotic cells that carries genes separate from those on the main chromosome and capable of independent replication. Plasmids can be used as vectors in the transfer of DNA. See vector.

plasmodesmata, ***sing.*** **plasmodesma** (*plasm-*, formed, + *-desm-*, to bind, + *-ata*, characterized by having): cytoplasmic strands that extend through pores in cell walls and connect the cytoplasms of adjacent cells.

plasmodium (*plasm-*, formed, + *-ium*, region): a membrane-bound, multinucleated mass of protoplasm lacking a cell wall and therefore often considered "naked." A stage in the life cycle of Myxomycota ("slime molds").

plasmogamy (*plasm-*, formed, + *-gam-*, union, + *-y*, process of): the fusion of the cytoplasms of two gametes prior to the fusion of their nuclei, resulting in a dikaryotic condition.

plasmolysis (*plasm-*, formed, + *-lys-*, to dissolve): a process in which the cell membrane pulls away from the cell wall resulting in the shrinkage of the cytoplasm; caused by a loss of water from the cell after being placed in a hypertonic solution.

plastid (*plast-*, membrane, + *-id*, tending to): a membrane-bound organelle functioning as a site for photosynthesis (chloroplasts), starch storage (amyloplasts and leucoplasts), or pigment storage (chromoplasts).

platelet (*plat-*, flat, + *-let*, small): a cell fragment found in blood involved in the process of blood clotting. Also referred to as a thrombocyte.

plectostele (*plect-*, twisted, + *-stel-*, a pillar): a modified form of actinostele being deeply fissured in cross section; common in some primitive vascular plants such as the club mosses.

pleiotropy (pleiotrophic gene) (*plei-*, many, + *-trop*, to turn or change + *-y*, process of): the ability of a single gene to influence or control several different phenotypic characteristics.

pleura (*pleur-*, side): a membrane that lines the thoracic cavity and covers the lungs and diaphram.

plexus (*plex-*, network, + *-us*, thing): a network of nerves, blood vessels, or lymphatic vessels.

ploidy (*ploid-*, multiple of, + *-y*, state of): a genetic term pertaining to chromosome number; such as haploid, diploid, polyploid, etc.

plumule (*plum-*, feather, + *-ule*, small): the terminal bud of an embryo including the embryonic leaves; the young shoot above the cotyledons.

pneumatocyst (*pneum-*, air, + *-cyst*, bag): a hollow, gas-filled region of the thallus of brown algae used for flotation.

poikilotherm—see ectotherm

point mutation: a mutation that alters a single nucleotide within a gene; such as the mutation that causes sickle-cell anemia.

polar bodies (*pol-*, end of an axis, + *-ar*, pertaining to, + bodies): small nonfunctioning cells that form during the development of female gametes (oogenesis); of the four haploid cells that result from oogenesis, one develops into a functional egg cell, the remaining three become polar bodies.

polar covalent bond (*pol-*, end of an axis, + *-ar*, pertaining to, + *co-*, together, + *-val-*, to be strong, + *-ent*, having the quality of, + bond): a relatively strong chemical bond in which the electrons are shared unequally between two atoms resulting in the formation of a polar molecule. See polar molecule.

polar molecule (*pol-*, end of an axis, + *-ar*, pertaining to, + molecule): a molecule with an unequal distribution of electrons causing one part of the molecule to have a positive charge and the opposite part to have a negative charge; such as water.

polar nuclei (*pol-*, end of an axis, + *-ar*, pertaining to, + nuclei): two of the eight haploid nuclei of an angiosperm megagametophyte that migrate to the middle of the embryo sac and fuse with one sperm nucleus to form the triploid (3n) endosperm nucleus.

pollen (grain) (*pollen*, fine dust): the mature microgametophyte of seed plants consisting of a haploid tube nucleus and two haploid sperm nuclei.

pollen sac—see anther

pollen tube: the tube formed by the germinating pollen grain that transports the sperm nuclei into the ovule.

pollination (*pollen-*, fine dust, + *-ation*, the process of): the transfer of pollen from the microsporangium to the stigma of the pistil in angiosperms, or to the micropyle of the ovule in gymnopserms.

polygenic (inheritance) (*poly-*, many, + *-gen-*, origin, + *-ic*, pertaining to): a hereditary mechanism involving the interaction of many genes that determines one specific trait; such as weight or height. The many genes involved will often have an accumulative effect on the specific trait; also referred to as quantitative inheritance.

polymer (*poly-*, many, + *-mer*, part of): a long chain-like molecule composed of many "links" (monomers) of the same general type bonded together in a specific way; nucleic acids and proteins are polymers.

polymerization (*poly-*, many, + *-mer-*, part of, + *-ization*, the process of): the anabolic process of forming a polymer; usually controlled by a polymerase molecule.

polymorphism (*poly-*, many, + *-morph-*, form, + *-ism*, the process of): the existence of two or more phenotypically different forms of the same trait within a population; such as human blood types.

polyp (*polyp,* small growth): (1) the sessile stage in the life cycle of certain cnidarians; Hydra-like organisms. (2) a benign tumor.

polypeptide (*poly-,* many, + *-pep-,* to digest, + *-ide,* denoting a chemical compound): a polymer of amino acids linked together by peptide bonds.

polyploid (*poly-,* many, + *-ploid,* multiple of): a condition in which a cell or organism has more than two complete sets of chromosomes per nucleus.

polysaccharide (*poly-,* many, + *-sacchar-,* sugar, + *-ide,* denoting a chemical compound): a long chained carbohydrate made of three or more monosaccharides; such as starch, cellulose, and glycogen, all of which are made of hundreds of monosaccharide monomers.

polysome (polyribosome) (*poly-,* many, + *-som-,* body): a cluster of ribosomes connected to one molecule of mRNA during translation.

polyunsaturated fat (*poly-,* many, + *-un-,* not, + *-satur-,* to fill up, + fat): a type of lipid (fat) consisting of several fatty acids that do not have the maximum number of hydrogen atoms bonded to carbon, thus not saturated with hydrogen. See fatty acid. Opposite of saturated fat.

pome (*pomme,* apple): a fleshy, simple, nonsplitting fruit, the outer edible part of which is formed from tissues (the hypanthium) surrounding the ovary. The mature ovary forms the "core" of the pome.

population (*popul-,* people, + *-ation,* process of): a group of organisms of the same species inhabiting a specific area and freely interbreeding with one another.

portal system (*port-,* gate, + *-al,* pertaining to, + system): a circulatory pathway joining two capillary beds; such as the hepatic portal system.

posterior (*poster-,* behind, + *-or,* state of): the rear or tail end of an organism. Opposite of anterior.

potential energy: stored energy; measured in joules. Opposite of kinetic energy.

preadaptation (*pre-,* before, + *-adapt-,* to fit, + *-ation,* the process of): an evolutionary change in an existing structure or behavior enabling the structure or behavior to have a different function.

prebiotic (*pre-,* before, + *-bi-,* life, + *-tic,* pertaining to the process of): pertaining to the process of chemical evolution that occurred prior to the evolution of life.

predation (*pred-,* to prey upon, + *-ation,* the process of): an ecological relationship in which one organism kills and eats another.

prehensile (*prehend-,* to seize, + *-ile,* having the character of): adapted for grasping; usually in reference to a prehensile tail common to some monkeys.

pressure flow hypothesis—see mass flow hypothesis

primary consumer—see consumer

primary germ layers—see germ layer

primary growth (*prim-,* first, + *-ary,* place for, + growth): growth originating in the shoot and root meristem providing length to the plant. Opposite of secondary growth, which provides girth (width). See secondary growth.

primary plant body (*prim-,* first, + *-ary,* place for, + plant body): all regions of the plant that develop from the apical meristems and their meristematic derivatives.

primary producer—see autotroph

primary protein structure (*prim-*, first, + *-ary,* place for, + *prote-*, primary, + *-in,* chemical substance, + structure): the first structural level of protein complexity; associated with and determined by the specific amino acid sequence of the protein.

primary succession (*prim-*, first, + *-ary,* place for, + *success-*, to follow, + *-ion,* process of): the gradual process of establishing life in an area previously devoid of all life forms, as on bare rock. See secondary succession.

primate (*prim-*, first, + *-ate,* characterized by having): a member of the taxonomic order Primates; humans are classified in this order.

primordium, *pl.* primordia (*primord-*, beginning, + *-ium,* structure): an organ or cell in its earliest stage of development; such as a leaf primordium.

prion (a contraction for a protein-like infectious agent): a subviral pathogen consisting of only a glycoprotein molecule; believed to be linked to several diseases of the Central Nervous System in humans, including kuru.

procambium (*pro-*, before, + *-camb-*, exchange, + *-ium,* region, part): the primary meristem of roots and shoots that gives rise to xylem and phloem.

producer—see autotroph

progeny (*pro-*, before, + *-gen-*, origin, + *-y,* process of): the offspring produced by parents.

progesterone (*pro-*, before, + *-gest-*, to carry): a female sex hormone produced in the ovaries and placenta; responsible for stimulating the growth and maintenance of the uterus during pregnancy.

prokaryotic (procaryotic) (*pro-*, before, + *-kary-*, nut, nucleus, + *-ic,* pertaining to): pertaining to an organism lacking membrane-bound organelles; such as bacteria and blue-green algae. Opposite of eukaryotic

promoter (*promo-*, to move forward): a specific sequence of DNA that serves as a binding site for RNA polymerase.

prophage (*pro-*, before, + *-phag-*, to eat): a segment of DNA from a bacteriophage (bacterial virus) that has been incorporated into a host chromosome.

prophase (*pro-*, before, + *-phase,* a stage): the first stage of mitosis and meiosis during which chromatin begins to condense into visible chromosomes.

proprioceptor (*propri-*, one's own, + *-cept-*, to take): a sensory receptor deep within tissue that senses movement and body position; especially abundant in striated muscle tissue.

prop roots: adventitious roots arising from the stem above the ground that help in supporting the plant; also referred to as buttress roots. Common in monocots, such as corn.

prostaglandins (*prostat-*, one who stands before, + *-gland-*, acorn, + *-in,* chemical substance): a group of powerful fatty acid hormones involved in regulating blood clotting, stimulating smooth muscle contractions, and various other physiological functions; believed to be produced by every cell in the human body.

protease (*prote-*, primary, + *-ase,* enzyme): an enzyme that digests proteins; such as peptidase.

protein (*prote-*, primary, + *-in*, chemical substance): a 3–D macromolecule made of amino acid subunits; a primary constituent of all cells.

proteinoid microsphere (*prote-*, primary, + *-in-*, chemical substance, + *-oid*, resembling, + *micr-*, small, + *-sphere*, round): a spherical arrangement of protein-like polymers formed under laboratory conditions by heating a mixture of dry amino acids and water; important as a model for ancestral cell types.

prothallium (*pro-*, prior to, + *-thall-*, a young shoot, + *-ium*, region): the independent heart-shaped gametophyte of ferns and club mosses. Also known as a prothallus.

protocell (*prot-*, primary, + cell): a hypothetical precursor to true cells. See liposome, protenoid microsphere, and coacervates.

protoderm (*prot-*, primary, + *-derm*, skin): the primary meristem of roots and shoots that gives rise to the epidermis.

proton (*prot-*, primary, + *-on*, a particle): a primary subatomic particle with a positive charge; found in the nucleus of an atom along with one or more neutrons.

proton pump: a form of active transport that generates a membrane potential by forcing hydrogen ions (protons) out of the cell; exists within the cell membrane and involves several enzymatic proteins.

protonema (*prot-*, primary, + *-nem-*, a thread): the early filamentous growth of the gametophyte in moss and ferns; the precursor of the leafy gametophyte.

protoplasm (*prot-*, primary, + *-plasm*, formed): all the living material within the cell membrane; a somewhat outdated term for living material.

protoplast (*prot-*, primary, + *-plast*, membrane): a plant or bacterial cell with its cell wall removed; often produced artifically by enzymatically digesting away the cell wall.

protostele (*prot-*, primary, + *-stel-*, a pillar): the simplest type of stele in which a solid core of xylem is surrounded by phloem. Common in dicot roots and the stems of primitive vascular plants.

protostome (*prot-*, primary, + *-stom-*, mouth): an animal in which the mouth develops from or near the blastopore and the anus develops from a secondary opening, and in which cleavage is determinate. One of two evolutionary lines of coelomate animals that includes mollusks, annelids, and arthropods. Opposite of deuterostome.

protractor (*pro-*, prior to, + *-tract-*, pull out, + *-or*, result of the act of): a muscle that extends or pulls a structure away from the body. Opposite of retractor.

proximal (*proxim-*, nearest, + *-al*, pertaining to): pertaining to a structure located near the origin or point of attachment. Opposite of distal.

pseudocoelom (*pseudo-*, false, + *-coel-*, cavity): a "false" fluid-filled body cavity; false because it is not completely lined with mesoderm; characteristic of the nematodes.

pseudopod (pseudopodium) (*pseudo-*, false, + *-pod*, foot): a temporary cytoplasmic extension from any ameboid cell such as the protozoan *Ameoba* or phagocytic white blood cells; used for locomotion, or for taking in food or foreign material.

pseudoscience (*pseudo-*, false, + *-scien-*, knowledge): a scientific-sounding idea that does not meet the criteria of the scientific method. Astrology is a pseudoscience because controlled experimental data do not exist. See scientific method.

ptyalin (*pty-*, to split, + *-in*, chemical substance): an enzyme found in saliva that breaks down starch into maltose. See amylase.

pubescence (*pubes-*, to grow hair, + *-ence*, the condition of): a downy or hairy covering.

pulmonary (*pulmon-*, lung, + *-ary*, place for): pertaining to the lungs; such as the pulmonary artery.

pulvinus (*pulvin-*, cushion, + *-us*, thing): an enlargement at the base of petioles from which leaf movement is controlled through changes in turgor pressure.

pump—see active transport

punctuated equilibrium: a modern theory of evolutionary change that suggests the rate of change is not slow and gradual (as proposed by Darwin and others) but rather one proceeding in radical "starts-and-stops" over relatively short periods of time. Long periods of no change (equilibrium) are interrupted (punctuated) by rapid bursts of speciation. Opposite of gradualism.

pupa (*pupa-*, doll, girl): an immobile developmental stage in some insects that may be protected within a cocoon; occurs between the larval and adult stage.

purine (*pur-*, pure, + *-ine*, having the character of): one of two classes of nitrogenous bases found in nucleotides (DNA, RNA, ATP, and NAD) made of two rings of carbon and nitrogen atoms; includes adenine and guanine. The other class is pyrimidine.

pyramid (ecological)—see ecological pyramid

pyrenoid (*pyren-*, the stone of a fruit, + *-oid*, resembling): a specialized structure within the chloroplasts of certain green algae where glucose is converted into starch.

pyrimidine (altered form of purine): one of two classes of nitrogenous bases found in nucleotides made of a single ring of carbon and nitrogen atoms; includes thymine, cytosine, and uracil. The other class is purine.

Q

Q_{10}—see temperature coefficient

quaternary protein structure: the fourth and final structural level of protein complexity; associated with and determined by the specific 3–D arrangement of the protein's polypeptide subunits.

quiescent center (*quies-*, to become quiet, + *-ent*, having the quality of, + center): a region in the apical meristem in which there is relatively little mitotic activity.

R

r-selection (r-strategist) (from the r value in the logistic equation): a survival strategy characterized by populations whose members mature relatively quickly, and produce many offspring that receive little or no parental care. Few of the offspring survive to reproduce; such as insects. Opposite of K-selection. See logistic growth (logistic equation).

raceme *(racem-,* cluster of berries): an inflorescence in which each flower is borne on a short stem (pedicle), and all are approximately the same length.

rachis *(rach-,* backbone): the main axis or midrib of a compound leaf or a fern frond.

radial cleavage *(radi-,* radiating, + *-al,* pertaining to, + *cleav-,* to divide, + *-age,* collection of): a type of early animal embryonic development in which the early cleavage planes are symmetrical to the polar axis resulting in straight rows of cells extending from pole to pole; characteristic of deuterostomes. See indeterminate cleavage.

radial section: a longitudinal section cut parallel to the radius of a cylindrical organ, such as a stem or root.

radial symmetry *(radi-,* radiating, + *-al,* pertaining to, + *symmet-,* measured together): (1) a body plan in which structures (tentacles for example) radiate outward from a central axis and, when viewed from above or below, looks circular; characteristic of cnidarians, adult echinoderms, and some sponges. (2) a flower that can be divided into two equal halves in more than one longitudinal plane. Also known as actinomorphic, or regular. Opposite of zygomorphic.

radicle *(radicl-,* a small root): an embryonic root.

radioactive *(radi-,* radiating, + active): referring to certain substances that give off energy in the form of particles and/or waves; characteristic of elements with unstable isotopes.

radioactive dating *(radi-,* radiating, + active, + dating): a method using the half-life of certain radioactive isotopes, such as carbon-14 and potassium-40, to date rocks and fossils; also referred to as carbon dating.

raphe *(raph-,* seam): (1) a longitudinal groove in the cell wall of some diatoms. (2) a ridge on some seeds formed by the stalk of the ovule.

ray: rows of parenchyma cells that extend radially through secondary xylem and phloem and that provide lateral conduction between the pith and cortex.

ray flower (floret): a flattened, irregular, ray-shaped flower on the margins of most composite inflorescences (heads) surrounding the disk flowers.

receptacle *(receptac-,* a little vessel): a fertile area on which reproductive organs develop; usually in reference to the part of the floral axis supporting the flowers.

receptor *(recept-,* to receive, + *-or,* state of): (1) a sensory cell or organ capable of detecting and responding to environmental stimuli, such as temperature receptors. (2) a molecule on the surface of cells that combines with specific molecules, such as hormones.

recessive (allele) (*recess-*, receding, + *-ive,* tending to, + *-allel-,* of one another): an allele (gene) that is expressed only when the genotype is homozygous (aa); the allele that is not expressed in a heterozygous condition. See dominant (allele).

recombinant DNA (*re-*, again, + *-combin-,* together, + *-ant,* that which, + DNA): DNA produced by combining segments of DNA from two or more different sources, such as viruses and bacteria; usually done under laboratory conditions. See biotechnology and genetic engineering.

red marrow: the tissue located in the spaces within the heads of long bones; responsible for producing blood cells.

redox reaction—see oxidation reduction reaction

reduction (*reduct-*, reduced, + *-ion,* the process of): the gain of one or more electrons by an atom or molecule (in reference to inorganic compounds), and the gain of one or more hydrogen ions (in reference to organic compounds); opposite of oxidation. Oxidation and reduction occur simultaneously with the electron lost by one molecule being readily taken up by another molecule

reduction division—see meiosis

refractory period: a relatively short period of time immediately after an action potential in which the nerve cell cannot respond to another stimulus.

regeneration (*re-*, again, + *-gen-*, origin, + *-ation,* the process of): the process of growing a new body part to replace a lost or diseased one; characteristic of crustaceans and other invertebrates.

regular flower—see radial symmetry

renal (*ren-*, kidney, + *-al,* pertaining to): pertaining to the kidney; such as the renal vein.

replication (*replic-*, to answer to, + *-ation,* the process of): the duplication of DNA from pre-existing molecules; occurs during interphase of the cell cycle.

repressor (*repres-*, to keep back, + *-or,* state of): usually in reference to a protein that blocks the expression of one or more genes.

reproductive isolation: the inability of different species to breed and produce fertile offspring.

resin (from *rhetine,* a yellowish liquid, + *-in,* chemical substance): an insoluable carbohydrate synthesized by certain plants and used as a protective mechanism against insect infestation and wood decay; commonly referred to as sap.

resolution (resolving power) (*resol-*, to unbind, + *-tion,* the process of): usually in reference to microscopy and the ability of a lens to distinguish two points as two distinct points; a measure of clarity.

respiration (*respir-*, to breathe, + *-ation,* the process of): the process of gaseous exchange (usually oxygen and carbon dioxide) between an organism and the environment. See cellular respiration.

resting potential: the polarity of the membrane of muscle cells and nerve cells at rest, with the inside of the cell being more negative than the outside. Opposite of action potential.

restriction enzymes: enzymes that cleave DNA molecules at specific points along the double helix; usually associated with the degradation of foreign DNA.

reticulate-net venation (*reti,* network, + *-ic-,* pertaining to, + *-ule,* little , + *-ate,* characterized by having, + *ven-,* vein, + *-ation,* the process of): an arrangement of leaf veins resembling a net. Also referred to as net venation.

reticulum (*ret-,* network, + *-ic-,* pertaining to, + *-ul,* little, + *-um,* structure): a fine network of filaments or filament-like structures within cells; such as endoplasmic reticulum.

retinol (*retina,* + *-ol,* denoting an alcohol)—**see vitamin A**

retractor (*retract-,* to pull back, + *-or,* result of the act of): a muscle that withdraws or pulls a structure toward the body. Opposite of protractor.

retrovirus (*retro-,* backward, + *-vir-,* poisonous slime, + *-us,* thing): a large group of cancer-causing RNA viruses that reproduce by transcribing its RNA into DNA.

Rh factors (named after the Rhesus monkey from which they were first isolated): specific antigens on the surface of red blood cells. Also known as Rhesus factors.

rhizoid (*rhiz-,* root, + *-oid,* resembling): an absorptive root-like filament lacking vascular tissue; common to the ventral surface of fern gametophytes.

rhizome (*rhiz-,* root, + *-ome,* mass): a horizontal underground stem.

riboflavin—see vitamin B complex

ribonucleic acid (RNA) (ribose, a pentose sugar, + *-nucle-,* nucleus, kernel, + *-ic,* pertaining to, + acid): a nucleic acid that functions mainly in the process of protein synthesis; composed of a single strand of nucleotides. RNA is also the genetic material of a large number of viruses. See messenger RNA, transfer RNA, and ribosomal RNA.

ribosomal RNA (rRNA) (ribose, a pentose sugar, + *-som-,* body, + *-al,* pertaining to): a type of ribonucleic acid (RNA) that, along with specific proteins, makes up ribosomes; transcribed from DNA found in the nucleolus. The most abundant type of RNA.

ribosomes (ribose, a pentose sugar, + *-som-,* body): minute cytoplasmic organelles often bound to endoplasmic reticulum composed of ribosomal RNA and proteins; the site of protein synthesis in eukaryotic cells.

ribozyme (ribose, a pentose sugar, + *-zym-,* ferment, enzyme): an enzymatic type of RNA used to produce new molecules of RNA.

root: an underground organ typical of most vascular plants that functions to anchor the plant and absorb and conduct water and minerals.

root cap: a protective tissue covering the root apical meristem.

root hair: an outgrowth from a single epidermal root cell that greatly increases the total absorptive surface area of the root system.

root pressure: a positive pressure in root xylem cells developed by the roots as a result of osmosis; causes guttation of water in leaves and cut stems.

rostrum (*rostr-,* beak, + *-um,* structure): the beak of birds or any beak-like process, as in certain arthropods.

rough ER—see endoplasmic retriculum

ruminants (*rumin-*, chew, + *-ant*, that which): animals with large multi-chambered stomachs specialized for an herbivorous diet and that chew cud; such as cattle and sheep.

S

S phase: the synthesis (S) phase of the cell cycle during which DNA is replicated; occurs prior to mitosis. See cell cycle.

sagittal section (*sagitt-*, arrow, + *-al*, pertaining to, + section): a section or division made in the median plane of an organism.

samara (*samara*, seed of the elm): a simple, dry, nonsplitting fruit containing one or two seeds; the pericarp is modified into wing-like extensions.

sap: (1) the fluid within xylem or phloem cells. (2) a common term improperly applied to resin and latex.

saprophytic (*sapr-*, rotten, + *-phyt-*, plant, + *-ic*, pertaining to): any plant-like organism that feeds off of and derives its energy from a dead organic source; such as most fungi and bacteria; may also be referred to as saprobic or saprozoic.

sapwood: secondary xylem in stems that conducts water; often light in color and surrounding the heartwood. See heartwood.

sarcolemma (*sarc-*, muscle, + *-lemma*, sheath): the specialized cell membrane of a muscle cell capable of propagating action potentials.

sarcomere (*sacr-*, muscle, + *-mer-*, a part of): the functional contractile unit of striated muscle cells located between adjacent Z-lines.

sarcoplasmic reticulum (*sacro-*, muscle, + *-plasm-*, formed, + *-ic*, pertaining to, + *reti-*, network, + *-ule*, little, + *-um*, structure): an extensive network of membranes and tubules within and surrounding muscle cells that stores and releases calcium. The calcium initiates muscle contraction.

saturated fat (*satur-*, to fill up, + *-ate*, characterized by having, + fat): a type of lipid (fat) consisting of several fatty acids that have the maximum number of hydrogen atoms bonded to carbon, thus saturated with hydrogen. See fatty acid. Opposite of polyunsaturated fat.

schizocarp (*schiz-*, split, + *-carp*, fruit): a dry, nonsplitting fruit containing one or more united carpels that split at maturity into separate one-seeded sections.

schizocoel (*schiz-*, split, + *-coel*, cavity): a body cavity formed by the splitting of embryonic mesoderm into two distinct layers; characteristic of protostomes.

Schwann cell (named after the German anatomist, Th. Schwann): a uninucleate cell surrounding certain nerve fibers; responsible for producing the myelin sheath.

scientific method: the logical, orderly arranged approach used to study nature; usually consisting of the following steps: information gathering -> hypothesis formulation -> prediction -> experimentation and analysis -> conclusion (explanation).

scientific name—see binomial system (of nomenclature)

sclereid (*scler-*, hard, + *-id*, tending to): a sclerenchyma cell possessing a thick, lignified secondary cell wall with numerous pits; may not be living at maturity. Also known as a stone cell.

sclerenchyma (*scler-*, hard, + *-en-*, in, + *-chym-*, infusion): a plant tissue composed of cells with thick, lignified secondary walls; generally dead at maturity and used for support; includes fibers and sclereids.

sclerosis (*scler-*, hard, + *-sis*, the process of): the excessive hardening of tissue due to the accumulation of fibrous connective tissue or other material. See atherosclerosis.

sclerotium (*scler-*, hard, + *-ium*, region): a hard, dormant, plasmodial stage in the life cycle of Myxomycota (''slime molds'').

scutellum (*scutell-*, a small shield, + *-um*, structure): (1) the single cotyledon of grass embryos specialized for absorbing nutrients from the endosperm. (2) any shield-like structure.

seaweed: a common term for large algae; usually in reference to brown and red algae.

sebaceous glands (*seb-*, grease, + *-aceous*, pertaining to, + gland): dermal glands that secrete an oil called sebum that is used to lubricate skin and hair.

secondary consumer—see consumer

secondary growth: growth originating from lateral meristems (the cork and vascular cambiums); provides girth. Opposite of primary growth that provides length. See primary growth.

secondary plant body: all tissues derived from the cork and vascular cambiums; includes secondary xylem, secondary phloem, and periderm.

secondary protein structure: the second structural level of protein complexity; usually in the form of a simple helix or pleated sheet of molecules; determined by the spontaneous folding of polypeptide chains. See alpha (α) helix and beta (β) pleated sheet.

secondary succession (*secund-*, second, + *-ary*, place for, + *success-*, to follow, + *-ion*, process of): the gradual process by which plant life becomes established on developed soil, often following a physical disturbance such as a fire; a process following primary succession. See primary succession.

secondary xylem (*secund-*, second, + *-ary*, place for, + *xyl-*, wood): xylem tissue derived from the vascular cambium; provides girth and strength to woody plants. Also referred to as wood.

secretion (*secret-*, to sever, separation, + *-ion*, process of): a glandular process whereby specific substances are separated from the blood and formed into new substances, such as milk, oil or bile.

sedentary (*sed-*, to sit, + *-ary*, pertaining to): usually in reference to an organism that is not free-living but permanently attached to one location, such as coral and sponges; also referred to as sessile. Opposite of free-living.

seed: a mature ovule consisting of an embryo, stored food such as endosperm, and a protective covering (seed coat).

segmentation (*segment-*, piece, + *-ation*, the process of): a body plan consisting of a series of segments that are more or less alike; associated with the evolution of specialized body parts and increased body size; also referred to as metamerism.

segregation (Mendel's first law): a fundamental principle of genetics in which two alleles segregate (separate) from each other during meiosis. Each resulting gamete has an equal chance of receiving either allele.

semipermeable (selectively permeable) (*semi-*, half, + *-perm-*, to pass through, + *-able*, tending to): the physical and chemical property of biological membranes that allows some substances (water and small ions) to pass through more easily than other substances (large molecules). Also referred to as differential permeability.

senescence (*senesc-*, to grow old, + *-ence*, the condition of): an aging process characterized by a decrease in cell division and the gradual loss of some metabolic functions, such as photosynthesis.

sensory neuron (*sens-*, to perceive, + *-ory*, related to, + neuron): nerve cells that receive sensory impulses, both external and internal, and carry them to the CNS.

sepal (*sep-*, covering, + *-al*, pertaining to): the outermost, sterile, leaf-like covering of a flower; usually green in color. Collectively, the sepals are called the calyx.

septum, ***pl.*** **septa** (*sept-*, partition, + *-um*, structure): a cross wall that develops in the hyphae of certain fungi. Cells with septa are said to be septate.

serum (*ser-*, the watery parts of fluids, whey, + *-um*, structure): any biological fluid that has been separated from its cellular and particulate components; usually in reference to the liquid that separates from blood after clotting.

sessile (*sess-*, without a stem, + *-ile*, having the character of): (1) usually in reference to leaves lacking a petiole or to a flower lacking a pedicle. Opposite of petiolate. (2) also in reference to an animal that is not free-living, such as coral.

seta, ***pl.*** **setae** (*set-*, a bristle): (1) a chitinous bristle located on the integument of many annelids and arthropods used for locomotion and defense. (2) the stalk that supports moss spore capsules.

sex chromosomes: chromosomes that carry genes involved in sex determination; an X (female) or Y (male) chromosome (or equivalent). Opposite of autosomes.

sex-linked trait (sex linkage): a genetic trait, unrelated to sexual traits, determined by genes located on one sex chromosome but not the other. Hemophilia and color blindness are human sex-linked traits.

sexual dimorphism (sexual, + *di-*, two, + *-morph-*, shape, + *-ism*, process of): pertaining to organisms possessing two distinct forms (male and female individuals) based on secondary sexual characteristics.

Sexually Transmitted Diseases (STD's): infections that are transmitted through sexual contact. The infectious agent may be viral (AIDS), bacterial (gonorrhea), or fungal (yeast).

sexual reproduction: any reproductive process involving the union of a sperm cell and egg cell; reproduction involving meiosis and fertilization. Opposite of asexual reproduction.

sheath: a general term for any covering external to a cell or organ; such as the gelatinous sheath surrounding blue-green algal cells, or the flattened base of grass leaves that wrap around the stem.

shoot: the aerial portion of a plant including the stem and its appendages (leaves, fruit, flowers, etc.).

short-day plants: plants that flower in reponse to decreasing day lengths (usually less than 12 hours) and a corresponding increase in night length. Opposite of long-day plants.

shrub (*shrubbe,* brushwood): a woody perennial with several stems arising from or near the ground; usually short without strong apical dominance.

sieve cell: the conducting phloem cell in primitive vascular plants; usually with tapered ends lacking sieve plates.

sieve (tube) element: one conducting cell in a series making up a sieve tube; primarily in angiosperms.

sieve plate: the porous end wall of a sieve tube element.

sieve tube: a series of sieve tube elements connected end-to-end and interconnected with a sieve plate.

silique (*siliqu-,* a pod): a simple, two-celled fruit characteristic of the mustard family; similar to a pea pod.

simple fruit: a fruit derived from a single carpel or several fused carpels of a single flower.

simple leaf: a leaf having a single blade that may or may not be lobed.

sink: a metabolically active region in plants to which food (carbohydrates) and hormones are transported by the phloem; such as an apical meristem. Opposite of source.

sinus (*sin-,* hollow, + *-us,* thing): any cavity or space within tissue, such as the air cavities in the carnial bones.

siphonostele (*siph-,* pipe, + *-stel-,* pillar): an arrangement of vascular tissue in which a cylinder of xylem and phloem surrounds a central pith.

sister chromatids—see chromatid

skeletal muscle (*skelet,* a dried hard body, + *-al,* pertaining to, + muscle): muscle tissue characterized by being striated, voluntary, and usually attached to and responsible for moving bones; also referred to as striated muscle or voluntary muscle.

sliding-filament theory: one of the currently accepted explanations for muscle contraction that states actin (thin) filaments slide across myosin (thick) filaments.

slime mold: a common term for fungal-like organisms classified in the Divisions Myxomycota and Acrasiomycota.

smooth ER—see endoplasmic reticulum

smooth muscle: muscle tissue characterized by not exhibiting a distinct banding pattern of contractile proteins (therefore smooth in appearance); involuntary, and found lining the walls of the digestive tract and arteries.

sodium-potassium pump: a special type of active transport mechanism located in the cell membrane of animal cells that transports sodium out of and potassium into the cell against their concentration gradients with the aid of ATP; essential in the conduction of nerve impulses.

solute (*solu-*, dissolve): any substance that is dissolved in a solvent to form a solution.

solution (*solu-*, dissolve, + *-tion*, process of): any homogeneous liquid composed of two or more substances.

solvent (*solv-*, loosen, + *-ent*, performing the action of): any liquid that acts as a dissolving agent. Water is considered the "universal" biological solvent.

somatic cells (*som-*, body, + *-tic*, pertaining to, + cell): any body cell except those giving rise to gametes. Opposite of germ cells.

Somatic Nervous System (SNS) (*som-*, body, + *-tic*, pertaining to, + nervous system): a part of the Peripheral Nervous System that controls skeletal muscles; the voluntary nervous system. Opposite of the involuntary nervous system. See Peripheral Nervous System.

somites (*som-*, body, + *-ite*, belonging to): paired, block-like segments (metameres) of mesoderm tissue arranged longitudinally along each side of a chordate embryo; eventually differentiates into the vertebral column and dorsal muscles.

soredium, ***pl.*** **soredia** (*sorus-*, a heap, + *-ium*, region): an asexual reproductive unit of lichens made of several algal cells surrounded by fungal hyphae.

sorus, ***pl.*** **sori** (*sorus-*, a heap): a cluster of sporangia on fern fronds.

source: a metabolically active region in plants in which food (carbohydrates) and hormones are produced or stored; such as a leaf and fruit. Opposite of sink.

speciation (*speci-*, a kind, + *-ation*, the process of): the evolution of a new species; also referred to as macroevolution.

species (*speci-*, kind): members of a population who possess similar characteristics and freely interbreed to produce fertile offspring. See population.

sperm (*sperm-*, seed): flagellated, haploid male gametes; also referred to as spermatozoa.

spermatocyte (*spermat-*, seed, + *-cyte*, cell): a diploid cell that eventually produces haploid sperm cells.

sphincter (*sphing-*, to bind tightly): a circular muscle surrounding an opening of a hollow organ, the constriction of which closes the opening; such as the pyloric sphincter at the lower end of the stomach.

spike (*spic-*, head of grain): a simple, elongated, usually unbranched inflorescence bearing sessile flowers. Common in the grass family.

spindle fibers (apparatus): an intercellular structure visible in dividing cells composed of microtubules that provides the framework upon which chromosomes move during mitosis and meiosis; also referred to as mitotic spindle.

spine (*spin-*, thorn): (1) a leaf modified for protection; usually hard and pointed. (2) a sharp bone process. (3) the spinal column.

spiracles (*spira-*, to breathe, + *-cle*, little): (1) small external respiratory openings found in terrestrial arthropods. (2) external openings for the movement of water, found in sharks, rays, and skates.

spiral cleavage (*spir-*, coil, + *-al*, pertaining to, + *cleav-*, to cut, + *-age*, collection of): a type of early embryonic development in which the early cleavage planes form spiraling rows of cells extending from pole to pole; characteristic of protostomes; also referred to as alternating cleavage. See determinate cleavage.

spirochete (*spir-*, coil, + *-chaet-*, hair): a spiral or corkscrew-shaped bacterium.

spongy (mesophyll) layer: a layer of irregular-shaped parenchyma cells with large intercellular spaces; just above the lower epidermis in leaves. See mesophyll.

sporangiophore (*spor-*, spore, + *-angi-*, container, + *-phore*, to bear): an erect stalk on which sporangia produce spores; common to most fungi and some primitive plants.

sporangium, *pl.* sporangia (*spor-*, spore + *-angi-*, container, + *-um*, part, region): a spore-producing structure that may be unicellular as in some algae or multicellular as in mosses.

spore (endospore) (*spor-*, spore): (1) in bacteria, an asexual, often haploid, reproductive structure capable of long periods of dormancy, after which it will develop into a new organism genetically identical to the parent. (2) in plants, a haploid reproductive structure produced by the sporophyte by meiosis that develops into a gametophyte; usually unicellular.

spore mother cell (sporocyte): a diploid cell that divides by meiosis to produce four haploid spores or nuclei.

sporic meiosis (*spor-*, spore, + *-ic*, pertaining to, + *mei-*, reduction, + *-sis*, the process of): the production of spores by meiosis; common in some algae, mosses, and all vascular plants.

sporophyll (*spor-*, spore, + *-phyll*, leaf): a leaf or leaf-like structure bearing sporangia.

sporophyte (*spor-*, spore, + *-phyt-*, plant): the diploid phase of a plant life cycle that produces haploid spores by meiosis. Opposite of gametophyte.

stamen (*stam-*, thread): the male part of a flower that produces pollen; usually consisting of an anther supported by a filament.

staminate (*stam-*, thread, + *-ate*, characterized by having): referring to a unisexual flower having stamens but no pistil.

staphylococcus (*staphyl-*, a cluster of grapes, + *-cocc-*, a berry, + *-us*, thing): an arrangement of cells in grape-like clusters; usually pertaining to spherical (coccus) bacteria.

starch (*sterchen-*, to stiffen): a complex soluble polysaccharide composed of hundreds of glucose molecules bonded in a linear arrangement; the main food storage carbohydrate of plants.

start codon: a codon that marks the beginning of a specific gene sequence; also referred to as an initiator codon. Opposite of stop codon.

statocyst (*stat-*, standing still, + *-cyst,* cell): a sensory cell or organ of equilibrium containing one or more granules (statoliths) that stimulate sensory hairs when moved; used in a variety of animals to sense gravity and motion.

statoliths (*stat-*, standing still, + *-lith-*, stone): cytoplasmic starch granules that settle to the low point of a cell in relation to gravity; gravity sensors in plants.

stele (*stel-*, a pillar): the cylinders of vascular tissue (xylem and phloem) that run throughout vascular plants conducting water and food materials.

stem: the aerial portion of most vascular plants and anatomically similar organs below the ground (corms, bulbs, tubers, rhizomes).

steroid (*ster-*, hard, solid, + *-oid,* resembling): a lipid compound composed of four interconnected carbon rings bonded to several types of functional groups. Several types of steroids exist including cholesterol, progesterone, testosterone, and vitamin D. Anabolic steroids stimulate any anabolic reaction. Also referred to as sterol.

stigma (*stigm-*, spot): (1) the apical portion of the pistil in flowers onto which pollen grains adhere. (2) the light-sensitive "eye-spot" in certain unicellular flagellated organisms.

stipe (*stip*-stalk): a general term for a stalk-like support structure lacking vascular tissue; as in some Basidiomycetes ("mushrooms") and brown algae.

stolon (*stol-*, to shoot): a stem that grows horizontally along the surface of the ground; such as the stolon of a strawberry plant. Also known as a runner.

stoma (*stom-*, mouth, opening): (1) a pore or opening in the epidermis of leaves bordered by guard cells. (2) any mouth-like opening.

stomata (*stom-*, opening): a collective botanical term for the stoma plus the guard cells.

stop codon: a codon that marks the end of a specific gene sequence; also known as a terminator codon. Opposite of start codon.

streptococcus (*strept-*, twisted chain, + *-cocc-*, a berry, + *-us,* thing): an arrangement of cells resembling "beads-on-a-string"; usually pertaining to spherical (coccus) bacteria.

striated muscle (*stria-*, groove, furrow, + *-ate,* characterized by having, + muscle): muscle tissue exhibiting a distinct banding pattern caused by the regular arrangement of contractile proteins; cardiac and skeletal muscle are two types of striated muscle. See sacromere.

strobilus, ***pl.*** **strobili** (*strobil-*, a cone, + *-us,* thing): an elongated reproductive structure composed of modified, spore-bearing leaves (sporophylls); also called a cone.

stroma (*strom-*, bedding): the dense, viscous liquid that surrounds the grana and thylakoids within a chloroplast; contains enzymes used in carbon fixation. See photosynthesis.

style (*styl-*, a column): the elongated, central portion of the pistil between the ovary and stigma.

suberin (*suber-*, the cork oak, + *-in,* chemical substance): a wax-like lipid used as a water-proofing material in cork cells and in the Casparian strip of endodermal cells.

subsidiary cells (*sub-*, below, + *-sid-*, to settle, + *-ary*, pertaining to, + cells): morphologically distinct epidermal plant cells situated adjacent to guard cells that assist in the functioning of stomata; also referred to as auxiliary or accessory cells.

substrate (*substrat-*, strewn under): (1) a substance upon which an enzyme acts; any reactant in an enzyme mediated reaction. (2) the foundation to which an organism is attached.

substrate-level phosphorylation (*substrat-*, strewn under, + level, + *phosphor-*, bringing light, + *-yl-*, chemical radical, + -ation, the process of): the formation of ATP by directly transferring a phosphate group to ADP during glycolysis; light energy is not needed as in photophosphorylation.

succession—see ecological succession

succulent (*succ-*, juice, + *-ent*, within): usually in reference to a plant with fleshy, water-storing leaves or stems; characteristic of the cactus family.

sucrose (*sucr-*, sugar, + *-ose*, resembling): the most common disaccharide produced by plants; common cane or table sugar; $C_{12}H_{22}O_{11}$.

sugar: a common term that applies to monosaccharides and disaccharides.

superior ovary (*super-*, above, + *-or*, state of, + *ov-*, an egg, + *-ary*, place for): an ovary situated above the point of attachment of the stamens, petals, and sepals. Opposite of inferior ovary.

suppressor T cell: a type of T lymphocyte that stops the production of certain T and B lymphocytes when specific immune activities are no longer required; also functions in the maintenance of immune tolerance.

survivorship curve: a graph showing the number of organisms alive in different age groups of the population; a quantitative way to show age-specific mortality.

symbiont (*sym-*, together with, + *-bi-*, life, + *-ont*, individual): one partner in a symbiosis that benefits from the association. See symbiosis.

symbiosis (*sym-*, together, with, + *-bi-*, life, + *-sis*, the process of): an intimate association between two or more different organisms; includes mutualism, parasitism, and commensalism. The algal-fungal association that makes up a lichen is considered a symbiosis. See mutualism.

symmetry (*symmet-*, measured together): a body plan that is the same on each side of a specific axis.

sympathetic nervous system (*sympath-*, of like feelings, + *-ic*, pertaining to, + nervous system): pertaining to the portion of the Autonomic Nervous System that prepares the body for a fight-or-flight response; such as increasing blood flow and blood sugar levels. Opposite of the parasympathetic nervous system.

sympatric speciation (*sym-*, together with, + *-patr-*, native land, + *-ic*, pertaining to, + *speci-*, a kind, + *-ation*, the process of): the evolution of new species within the same geographical region as the parent species. Opposite of allopatric speciation.

symplast transport (*sym-*, together with, + *-plast*, membrane, + transport): the movement of materials via interconnected protoplasts and their plasmodesmata. Opposite of apoplast transport.

synapse (*synap-*, a union): any junction between two structures or cells; such as the junction between two neurons or between a neuron and an effector cell or organ.

synapsis (*synap-*, a union, + *-sis*, process of): the pairing up and eventual joining of replicated homologous chromosomes during prophase I of meiosis.

synecology (*syn-*, together + *-ec-*, dwelling + *-logy*, study of): the ecology of whole communities, involving structure, devolopment, and distribution. Opposite of autecology.

synergids (*synerg-*, work together, + *-id*, tending to): two sterile haploid cells on either side of the egg cell within the embryo sac of angiosperms.

syngamy—see fertilization

synthesis (reaction) (*syn-*, together, + *-the-*, to place, + *-sis*, the process of): a chemical reaction in which a new substance is produced from a number of simpler substances. See anabolism.

synthetic theory of evolution—see neo-Darwinism

systematics (*system-*, a composite whole, + *-tic*, pertaining to): the branch of biology dealing with the diversity and phylogenetic history of life; includes taxonomy (the naming of organisms) and classification (placing organisms in the correct phylogenetic position).

systole (*systol-*, a contraction): a period when heart muscle contracts and blood is forced out of the heart. Opposite of diastole.

T

T lymphocyte (T cell) (*lymph-*, a clear fluid, + *-cyte*, cell): a type of white blood cell that directly attacks and destroys foreign substances, cells, or tissues (cell-mediated immunity). The "T" refers to "thymus," a gland just above the heart in which these cells mature.

tactile (*tact-*, to touch, + *-ile*, pertaining to): pertaining to the sense of touch.

tagmosis (*tagm-*, arrangement, order, + *-sis*, the process of): the development of groups of body segments (metameres) into functionally distinct body regions (tagmata); such as the head, thorax, and abdomen regions of arthropods; also referred to as tagmatization.

tangential section (*tang-*, to touch, + *-al*, pertaining to, + section): a longitudinal section cut at right angles to the radius of the shoot, root, or other cylindrical structure.

tapetum (*tape-*, carpet, + *-um*, structure): a nutritive layer of tissue within a sporangium; usually in reference to a layer of cells lining pollen sacs (microsporangia).

taproot: the stout, tapering root formed from the radicle of the embryo. Also referred to as primary root.

taxis (*taxis*, ordered movement): a movement, either toward or away from a specific environmental stimulus, such as heat; usually in reference to simple organisms.

taxon, *pl.* **taxa** (*tax-*, arrangement): a category in a taxonomic system; such as kingdom, phylum, class, order, family, genus, species.

taxonomy (*tax-*, arrangement, + *-nomy,* the science of): the science of describing, naming, and classifying organisms.

tectum (*tectum,* roof): any roof-like covering over a cavity or structure; such as the dorsal covering in certain arthropods.

telencephalon (*tel-*, end, + *-encephal-*, brain): the anteriormost portion of the vertebrate forebrain containing the cerebrum and olfactory bulbs. The posterior portion of the forebrain is referred to as the diencephalon.

telophase (*tel-*, end, + *-phase,* a stage): the last stage of mitosis or meiosis during which new nuclei begin to form prior to cytokinesis. See cell cycle.

temperature coefficient (Q_{10}): the ratio of the rate of a reaction at a given temperature. The reaction rate of biological enzymes approximately doubles for every 10°C within a range of 0°C–40°C. Symbolized Q_{10}.

template: a mold or pattern used as a guide in the formation of a copy; usually in reference to the replication of DNA in which one strand of DNA acts as a template for the new complimentary strand.

temporal isolation (*temp-*, period of time, + *-al,* pertaining to, + isolation): a prezygotic isolating mechanism that prevents mating between two populations of similar organisms because they reproduce at different times of the day, season, or year.

tendon (*tend-*, to stretch): extremely dense and tough connective tissue that joins muscle to bone or muscle to muscle.

tendril (*tend-*, to stretch, + *-il* , little): a thread-like extension of a modified leaf or stem used for grasping objects; a means of mechanical support.

teratogen (*terat-*, monster, + *-gen,* origin): any biological, chemical, or physical agent that can cause a birth defect.

terminator codon: a codon found in messenger RNA that marks where protein synthesis ends. Opposite of initiator codon.

territorial (*territor-*, domain ,+ *-ial,* pertaining to): a defensive behavior in which animals closely protect a specific area needed for food and reproduction.

tertiary protein structure (*terti-*, three, + *-ary,* pertaining to, + protein structure): the third structural level of protein complexity; usually in the form of a complex molecular arrangement resulting spontaneously from further folding of the secondary structure of proteins; maintained by several types of interactions between adjacent polypeptide chains including hydrogen bonds, ionic bonds, and disulfide bonds.

test (*test,* shell): a hard shell or outer covering often impregnated with silica dioxide or calcium carbonate; such as the shells common to mollusks and all diatoms.

test cross: a genetic cross between an individual of an unknown genotype with a homozygous recessive individual (the "tester") to determine the genotype of the unknown. The resulting phenotypic ratio will determine the unknown genotype.

testis, *pl.* testes (*testis,* testicle): a male sex organ that produces sperm (male gametes) and sex hormones; may also be referred to as a testicle.

testosterone (*testis,* testicle, + *-ster* -, solid): a male sex horomone produced in the testis and responsible for maintaining secondary sexual characteristics, such as facial and body hair, deep voice, and specific bone and muscle characteristics.

tetanus (*tetan-,* stretched, + *-us,* thing): (1) a sustained contraction of skeletal muscle without relaxation resulting from a rapid succession of nerve impulses. (2) a disease caused by toxins produced by the bacterium *Clostridium tetani* growing in a wound, the symptoms of which include muscle spasms; also referred to as "lockjaw."

tetrad (*tetra-,* four): four chromatids that have come together as a result of the replication of paired homologous chromosomes during prophase I of meiosis. Also referred to as a bivalent.

tetraploid (*tetra-,* four + *-ploid,* mulitple of): a condition in which twice the normal number of chromosomes exist; symbolized 4n.

thallus (*thall-,* a young shoot, + *-us,* thing): a plant body that is not differentiated into true roots, stems, and leaves; usually in reference to algae and fungi.

theory (*theor-,* to look at, + *-y,* process of): a generalized statement based on repeated observations and experiments; a hypothesis that has been verified ("stood the test of time").

thermoperiodism (*therm-,* temperature, + *-period-,* a going around, + *-ism,* the process of): growth and development in response to differences in temperature.

thermoreceptor (*therm-,* heat, + *-recept-,* to receive, + *-or,* result of the act of): a sensory cell or organ capable of detecting and responding to a temperature stimulis, such as the sensory cells found in the skin of vertebrates.

thiamine—see vitamin B complex

thigmotropism (*thigm-,* touch, + *-trop-,* to turn, + *-ism,* the process of): a direction of growth in response to touch; grape tendrils are positively thigmotropic.

thoracic (*thorax,* breastplate, + *-ic,* pertaining to): pertaining to the chest; such as the thoracic cavity.

thorax (*thorax,* breastplate): (1) the upper portion of the vertebrate trunk containing the heart and lungs. (2) the portion of the arthropod body between the head and abdomen.

thorn: a stem modified for protection; usually hard and pointed.

thrombocyte (*thromb-,* clot, + *-cyt-,* cell)**—see platelet**

thylakoids (*thaylk-,* a bag, + *-oid,* resembling): flattened sac-like vesicles within chloroplasts in which photosynthetic pigments are arranged; the site of the light-dependent reactions and of photophosphorylation. Stacks of thylakoids form the grana.

thymine (*thym-,* emotions, + *-ine,* having the character of): one of four nitrogenous nucleotide bases that is a component of DNA; complimentary to adenine. See pyrimidine.

thymus (*thym-,* mind, + *-us,* thing, + gland): an endocrine gland located in the neck and chest region of young mammals; essential for the development of immunity and T cell differentiation.

thyroid gland (*thyr-*, an oblong shield, + *-oid*, resembling, + gland): an endocrine gland in the neck region of vertebrates that produces hormones (thyroxin) that regulate various metabolic rates.

tight junction: a type of intercellular junction between adjacent animal cells producing a tight seal (actual fusion of cell membranes), preventing material from leaking through spaces between cells. Intestinal epithelial cells possess tight junctions.

tissue (*text-*, to weave): a group of similar cells with a common structure and function.

tissue culture: a technique used to keep plant or animal tissues alive after removal from the organism, using an artifical growth medium.

tonicity (*ton-*, stretched, + *-ity*, state of): the elasticity of muscle, arteries, and other tissues; also referred to as tonus.

tonoplast (*ton-*, streched, + *-plast*, membrane): an inner membrane enclosing the central vacuole of mature plant cells.

totipotent (totipotency) (*toti-*, all, many, + *-potent*, to be powerful): the ability of certain cells to develop into a complete new organism. Cells that are totipotent are also undifferentiated and may be embryonic or meristematic in nature.

toxin (*tox-*, poison, + *-in*, chemical substance): a poisonous substance produced by a biological organism.

trace element—see micronutrients

tracheid(e) (*trache-*, windpipe): a lignified water-conducting xylem cell; usually dead at maturity.

tracheophyte (*trache-*, windpipe, + *-phyt-*, plant): an outdated term for plants with well-developed vascular tissue.

transcription (*trans-*, through, across, + *-scrib-*, to write, + *-tion*, the process of): the transfer of genetic information from DNA into messenger RNA (mRNA).

transduction (*trans-*, through, across, + *-duct-*, to lead, + *-tion*, the process of): the transfer of DNA fragments from one cell to another via a virus; usually in reference to bacterial cells.

transfer RNA (tRNA) (*trans-*, through, across, + *-fer-*, to carry): a type of ribonucleic acid (RNA) that transports individual amino acids to the site of protein synthesis on a ribosome.

transformation (*trans-*, through, across, + *-form-*, to shape, + *-ation*, the process of): (1) the process by which bacterial cells incorporate DNA from the external environment; usually in reference to a laboratory procedure. (2) the process by which a normal cell becomes a cancerous cell.

transfusion tissue (*trans-*, through, across, + *-fus-*, to pour, + *-ion*, the process of, + tissue): tissue composed of xylem cells and parenchyma cells surrounding the vascular bundles in gymnosperm leaves; used to conduct materials within the leaf.

translation (*trans-*, through, across, + *-lat-*, to carry, + *-ion*, the process of): the process of converting the information on a mRNA molecule into a sequence of amino acids and eventually into a protein.

translocation (*trans-*, through, across, + *-loc-*, place, + *-ation*, the process of): (1) the movement of substances within plants; usually in reference to the movement of phloem sap. (2) the process in which a portion of chromosome breaks off but reattaches to the same chromosome or another one.

transpiration (*trans-*, through, across, + *-spir-*, to breath, + *-ation*, the process of): the loss of water vapor from the aerial portions of plants; over 90% of the loss occurs through stomata, 10% through lenticels.

transpirational-pull theory: the proposed mechanism that states water is pulled up through the plant's xylem from the top of the plant rather than pushed up by root pressure; also referred to as the cohesion-adhesion-tension theory. See cohesion-adhesion-tension theory.

transport vesicles: relatively small membrane-bound cytoplasmic vesicles that move various substances between membrane systems, such as the movement of proteins between the endoplasmic reticulum and the Golgi apparatus.

transposon (*transpos-*, to reverse the order of, + *-on*, a particle): a segment of DNA that has the ability to move from one chromosome to another, or to different locations on the same chromosome resulting in genetic diversity; also referred to as a "jumping gene."

transverse section—see cross section

tree: a woody perennial with usually one stem or trunk.

trichome (*trich-*, hair, + *-ome*, mass): (1) an epidermal outgrowth such as the hair on leaves. (2) a chain of blue-green algal cells.

triglyceride (*tri-*, three, + *-glyc-*, sweet, sugar, + *-ide*, denoting a chemical compound): a common lipid composed of three fatty acids linked to a single glycerol molecule; the molecular component of animal fat and plant oil.

triplet code—see genetic code

triploblastic (*tri-*, three, + *-ploid-*, multiple of, + *-blast-*, sprout, bud, + *-ic*, pertaining to): pertaining to animals with three embryonic germ layers: ectoderm, mesoderm, and endoderm. Opposite of diploblastic.

triploid (*tri-*, three, + *-ploid*, mulitple of): usually a normal genetic condition in which a cell or organism has three sets of chromosomes per nucleus; such as the endosperm tissue of many angiosperm seeds; symbolized 3n.

trochophore (*troch-*, wheel, + *-phor-*, to bear): the free-swimming ciliated larval stage of mollusks and several types of marine worms.

trophic level (structure) (*troph-*, nutrition, + *-ic*, pertaining to, + level): a feeding level in an ecosystem; such as producers or consumers. A level in the flow of energy and biomass through an ecosystem.

trophoblast (*troph-*, nutrition, + *-blast*, bud, sprout): the outer layer of cells of a mammalian blastocyte that eventually gives rise to the chorion layer and placenta. See chorion.

tropism (*trop-*, to turn, + *-ism*, the process of): a growth response (movement) toward (positive) or away (negative) from an external stimulus. See phototropism.

tropomyosin (*trop-*, to turn, + *-myo-*, muscle, + *-in*, chemical substance): a globular protein that blocks the actin-binding sites on myosin, thereby preventing contraction of striated muscle; a component of the thick filaments in myofibrils.

tube cell: a cell within the pollen grains of seed plants which develops into the pollen tube.

tuber (*tuber-*, a swelling): an enlarged, fleshy, underground stem used to store food; such as the potato.

tubercle (*tuber-*, swelling, + *-cle*, little): a small rounded elevation or nodule on a bone or on the skin.

tubulin (*tubul-*, small tube, + *-in*, chemical substance): a globular protein that is the basic structural and functional unit of microtubules.

tumor (*tum-*, to swell, + *-or*, state of): a mass of cancerous tissue.

tunica (*tunic-*, cloak, covering): any thin membrane that covers or surrounds an organ; such as the outermost tissue layer of an artery, the tunica adventitia.

turgid (*turg-*, to swell, + *-id*, pertaining to): a firm condition of plant cells resulting from the uptake of water. See turgor (pressure). Opposite of flaccid.

turgor (pressure) (*turg-*, to swell, + *-or*, state of): the pressure exerted by the fluid contents on the inside of a plant cell wall. This pressure provides strength and rigidity to plants and plant cells. Cells with turgor pressure are said to be turgid.

type I herpes: a relatively common viral infection characterized by cold sores or fever blisters around the nose and mouth.

type II herpes: a highly contagious, sexually transmitted viral infection characterized by blisters on the genitals.

U

ulcer: an open sore usually located within mucous membranes or the skin; such as peptic ulcers located in the lining of the stomach.

ultrastructure (*ultra-*, beyond the normal, + structure): the level of cellular detail visible only with the electron microscope; typically anything less than 0.3 μm in diameter, such as mitochondria.

umbel (*umbell-*, a sunshade): an umbrella-shaped inflorescence in which pedicels radiate from a common point on the peduncle.

undifferentiated—see totipotent

undulipods (*undulat-*, risen like waves, + *-pod-*, foot): a recently proposed collective term for the cilia and flagella of eukaryotes.

ungulate (*ungul-*, hoof, + *-ate*, characterized by having): any hoofed mammal, such as a horse.

unicellular (*uni-*, one, + *-cellul-*, a small room, + *-ar*, pertaining to): an organism that exists as a single cell; such as a bacterium or protozoan.

uniramous (*uni-*, one, + *-ram-*, branch, + *-ous*, pertaining to): with reference to appendages with only one distinct branch. The opposite of biramous.

unisexual—see dioecious

unsaturated fat—see polyunsaturated fat

uracil (*ur-*, urine): one of four nitrogenous nucleotide bases that is a component of RNA; complimentary to adenine and replaces thymine in all types of RNA.

urea (*ur-*, urine): the primary nitrogenous waste product excreted in the urine of mammals; a water-soluble end product of protein metabolism.

ureter (from *ourein,* to urinate): one of the paired ducts that passes urine from the kidney to the bladder.

urethra (*ureth-*, canal): a single duct that passes urine from the bladder to the outside of the body; also serves to conduct semen in males.

uric acid (*ur-*, urine, + *-ic,* pertaining to, + acid): the primary nitrogenous waste product excreted by insects, birds, and reptiles; an insoluble end product of protein metabolism.

uterus (*uter-*, womb, + *-us,* thing): a highly muscular hollow organ in females in which the fetus develops; also referred to as the womb.

V

vaccine (*vacc-*, cow, + *-ine,* having the character of): a harmless, commercially produced antigenic solution that will produce an immune response to a specific pathogen when introduced into the body.

vacuole (*vacu-*, empty, + *-ole,* little): a membrane-bound organelle found within the cytoplasm of both plant and animal cells. Vacuoles are usually fluid-filled and function to regulate water concentration, but may also serve as a reservoir for food (food vacuoles) or enzymes (lysosomes).

valve (*valv-*, a folding door): (1) one half of a diatom frustule. (2) one half of a molluska shell.

variation (*vari-*, change, + *-ation*, the process of): anatomical or physiological differences within a population in response to environmental conditions, or based on genetic differences.

vascular (*vascul-*, a small vessel, + *-ar,* pertaining to): any plant or animal with conducting tissues or structures; such as xylem and phloem in plants, or veins and arteries in animals.

vascular bundle (*vascul-*, a small vessel, + *-ar,* pertaining to, + bundle): an elongated strand of xylem and phloem and, in some cases, supportive parenchyma tissue.

vascular cambium (*vascul-*, a small vessel, + *-ar,* pertaining to, + *camb-*, exchange, + *-ium,* region): a lateral meristem that produces secondary xylem (wood) toward the inside of the plant and secondary phloem toward the outside; typical of woody plants.

vascular tissue (system): one of three fundamental tissue systems in plants that gives rise to vascular and support tissues. See dermal tissue and ground tissue.

vasopressin—see antidiuretic hormone

vector (*vector,* carrier): (1) an agent that carries a parasite or pathogen (virus, bacteria, insect, or worm) from one host to another. (2) a portion of DNA into which a second DNA segment can be inserted (“spliced”) and incorporated into a cell. See plasmid.

vegetative (propagation or reproduction)—see asexual reproduction

vegetative body—see mycelium

vein (*ven-*, vein): (1) a thin-walled vessel that carries unoxygenated blood toward the heart. Opposite of artery. (2) an individual vascular trace in leaves.

venation (*ven-*, a vein, + *-ation*, the process of): the specific arrangement of veins in a leaf.

venereal disease (*vener-*, coitus, + *-eal*, pertaining to)—**see Sexually Transmitted Diseases**

ventral (*ventr-*, belly, + *-al*, pertaining to): pertaining to the front surface of an organ or animal that stands erect, such as man; or to the belly region of animals, such as snakes. Opposite of dorsal.

ventricle (*ventr-*, belly, + *-cle*, small): a cavity within an organ, such as the muscular chambers of the heart that pump blood, or the chambers within the brain.

venule (*ven-*, vein, + *-ule*, little): a small vein that takes blood from capillaries to veins. Opposite of arteriole.

vermiform appendix (*verm-*, worm, + *-form*, shape, + *append-*, an addition): in man, a blind, functionless, "worm-like" tube extending from the tip of the cecum of the large intestine; a remnant of the cecum in some mammals.

vernalization (*vernal-*, spring, + *-zation*, state of): the induction of flowering by means of a cold treatment; usually several degrees above freezing.

vesicle (*vesicle*, small bladder): a relatively small, sac-like, membrane-bound organelle within the cytoplasm used for the transport and storage of food, enzymes, and other organic compounds.

vessel (*vascul-*, a small vessel): (1) a multicellular tube composed of nonliving, lignified vessel elements with perforated end walls. The primary water-conducting system in angiosperms. (2) any duct or canal for transporting blood or lymph fluid in animals.

vessel element (member): one cell in a series making up a vessel which is unique to angiosperms.

vestigal (organ) (*vestig-*, trace, footprint, + *-al*, pertaining to): an evolutionary remnant of a structure that was once functional but has no known function at present; such as the vermiform appendix in man.

villus, *pl.* villi (*vill-*, tuft of hair, + *-us*, thing): a minute, hair-like projection on the membrane of cells lining the small intestine. Villi function to increase the absorptive surface area of intestinal cells.

viroid (*vir-*, poisonous slime, + *-oid*, resembling): a relatively short RNA molecule that can cause disease in plants and animals; also referred to as a "naked virus" since a protein coat surrounding the RNA is lacking. See virus.

virulent (*vir-*, poisonous slime, + *-ent*, having the quality of): possessing the ability of causing a disease.

virus (*vir-*, poisonous slime, + *-us*, thing): a noncellular, submicroscopic pathogen composed of a nucleic acid (DNA or RNA) core and a protein coat. Viruses can only reproduce if they infect a host cell.

viscera (*visc-*, internal organs): a collective term for all internal organs, particularly the organs in the abdominal cavity, such as the stomach and intestines.

vitamin (*vit-*, life, + *-amine*, of chemical origin): an organic compound required in small amounts for normal growth and development; usually not synthesized by the organism in which the vitamin is needed and must therefore be ingested preformed. Vitamins function primarily as coenzymes or cofactors of enzymes. See coenzyme.

vitamin A: a yellowish fat-soluble compound found primarily in green and yellow vegetables, and some fish liver oils; required for the development of bone and teeth, and normal vision. Also known as retinol.

vitamin B complex: a group of colorless water-soluble compounds found in a wide range of foods including yeasts, grains, fish, eggs, legumes, and fruits. Vitamin B_1 (thiamine) is a coenzyme precursor essential in carbohydrate metabolism; deficiency leads to beriberi in humans. Vitamin B_2 (riboflavin / vitamin G) is a coenzyme precursor essential in the metabolism of all major nutrients; deficiency leads to digestive problems. Vitamin B_6 (pyridoxine) is a coenzyme precursor essential in amino acid metabolism; deficiency leads to stunted growth. Vitamin B_{12} (cyanocobalamin) is a coenzyme precursor essential in the synthesis of DNA and development of red blood cells; deficiency leads to anemia. Other compounds within this complex include folic acid, biotin (vitamin H), and choline.

vitamin C: a colorless water-soluble compound found primarily in green vegetables and citrus fruits; required for the maintenance of connective tissues; deficiency leads to scurvy. Also known as ascorbic acid.

vitamin D: a colorless fat-soluble compound found primarily in cod liver oil and dairy products (also produced in the skin when exposed to sunlight); required for normal calcium and phosphorous metabolism, and bone development. Deficiency leads to abnormal skeletal development (rickets).

vitamin E: a colorless fat-soluble compound found primarily in cereal grains and green vegetables; required for the maintenance of cell membrane structure; deficiency leads to infertility, and muscular dystrophy.

vitamin K: a colorless fat-soluble compound found primarily in green vegetables and egg yolks; required for blood clot formation; deficiency leads to prolonged blood-clot formation. Deficiency is rare since it is manufactured by intestinal bacteria.

vitreous humor (*vitr-*, glass, + *-ous*, pertaining to, + *humor*, liquid): a body of clear viscous gel occupying the space between the lens and the retina of the eye; also referred to as the vitreous body.

viviparous (*viv-*, alive, + *-par-*, give birth to, + *-ous*, pertaining to): pertaining to an animal bearing "living" young that develop within the body of the mother. The offspring are born fully developed although sexually immature; such as mammals and many reptiles.

voluntary muscle—see skeletal muscle

W

water potential: a measure of the free energy of water in a cell or soil and therefore its tendency to move by osmosis; symbolized ψ.
wavelength: the distance between crests of waves measured in nanometres (nm) and symbolized (λ); usually in reference to wavelengths of light energy (electromagnetic energy).
wax: a plant lipid used as a waterproofing material. See cutin and suberin.
weed: any plant growing in a location where it is not desired.
white corpuscle—see leukocyte
white matter: nervous tissue possessing myelinated nerve fibers and therefore appearing glistening white; as opposed to gray matter.
whorl: a circle of leaves, sepals, or petals.
wild type: an organism with the normal, usually most common, phenotype.
wood: a common term for secondary xylem. See secondary xylem.

X

X-chromosome—see sex chromosomes
xanthophyll (*xanth-*, yellow, + *-phyll*, leaf): a yellow photosynthetic pigment. See carotenoid.
xerophyte (*xer-*, dry, + *-phyt-*, plant): a plant adapted for growth under dry conditions.
xylem (*xyl-*, wood): vascular tissue that conducts water and minerals from the roots to the rest of the plant; also functions in mechanical support.

Y

Y-chromosome—see sex chromosomes
yellow fat: the most common form of adipose tissue in animals containing very few mitochondria and therefore appearing yellow. Opposite of brown fat.
yellow marrow: the tissue located in the shaft of long bones containing a large amount of yellow fat; a primary site for fat storage. Opposite of red marrow.
yolk sac: one of the four extraembryonic membranes that supports the embryos of amniotes (mammals, birds, and reptiles). In reptiles and birds it performs a nutritive function; also referred to as the vitelline membrane.

Z

zeatin (*zea-*, a kind of grain, + *-in*, chemical substance): a natural cytokinin isolated from corn that controls cell division, bud growth, and senescence.
zooplankton (*zoo-*, an animal, + *-plank-*, drifting, + *-on*, a particle): a motile heterotroph present in plankton. See phytoplankton.
zoosporangium (*zoo-*, an animal, + *-spor-*, spore, + *-angi-*, container, + *-um*, structure): a cellular structure responsible for producing zoospores.
zoospore (*zoo-*, an animal, + *-spor-*, spore): a flagellated spore; common to primitive plants and algae.

zygomorphic (*zyg-*, paired together, + *-morph-*, shape, *-ic*, pertaining to): usually in reference to bilaterally symmetrical flowers. Opposite of actinomorphic. See bilateral symmetry.

zygospore (*zyg-*, paired together, + *-spor-*, spore): a thick-walled, resistant diploid spore resulting from the fusion of isogametes.

zygote (*zyg-*, paired together): a diploid (2n) cell resulting from the fusion of sperm and egg (or equivalent haploid cells); a fertilized egg.

zygotene (*zyg-*, paired together, + *-ten-*, to hold): the second stage of prophase I in meiosis during which synapsis of homologous chromosomes occurs.

zygotic meiosis (*zyg-*, paired together, + *-tic*, pertaining to, + *mei-*, reduction, + *-sis*, the process of): a situation in which meiosis occurs directly in the zygote; the zygote is the only diploid stage in the life cycle; common in some algae and fungi.

zymogen granules (*zym-*, enzyme, ferment, + *-gen-*, origin, + granules): membrane-bound organelles containing hydrolytic digestive enzymes in an inactive form (proenzymes); produced by pancreatic acinar cells.

5

Classification of Life

The following classification scheme has been modified to be practical and easy-to-use. It is based on the generally accepted "five kingdom system" of classification that is used in virtually all modern biology textbooks. It contains the major taxonomic categories (taxa) including the five kingdoms along with all major divisions (phyla are equivalent to divisions and are used in the kingdom Animalia), and selected classes along with a concise description of each. The coverage of all the taxa (kingdom, phylum or division, class, order, family, genus, species) here is far from complete and deals only with those taxa most often encountered by first-year students.

Many ways exist to classify living organisms, all of which, unfortunately for the student, are correct. Taxonomists meet yearly to modify their classification schemes of the estimated five million life forms, and this does not help the beginning student either! The classification scheme presented here is common to most first-year university/college biology texts.

One of the most useful features of this chapter is that a translation of the Greek and Latin roots, prefixes, and suffixes used to compose each taxa or level of classification is given. So, once you have seen that the meaning of the root word *-phyt-* means "plant," it will always mean plant or plant-like quality, regardless of whether the word represents a kingdom, division, or class!

Kingdom Monera (*mon-*, one, + *-er-*, connected with)

Single-celled (or colonial), microscopic prokaryotes characterized by being very simple in structure with no membrane-bound organelles; and either autotrophic (phototrophic or chemotrophic) or heterotrophic (saprophytic or parasitic). Reproduction is primarily asexual (binary fission or budding) although a primitive form of sexual reproduction does occur through conjugation and the exchange of plasmids. About 3000 species are classified within the kingdom Monera.

Division: Archaebacteria (*arch-*, ancient, first, + *-bacter-*, small rod): the most primitive forms of life on earth including three surviving subdivisions whose members live in extreme environments: the obligate anaerobe methanobacteria (methane producing bacteria), the halophiles ("salt-loving"), and the thermoacidophiles ("hot-acid loving").

Division: Eubacteria (*eu-*, true, + *-bacter-*, small rod): an extremely large group of unicelluar prokaryotes including all bacteria. Most are heterotrophic saprophytes or parasites. Essential decomposers.

Division: Cyanobacteria (*cyan-*, blue, + *-bacter-*, small rod): "blue-green algae"; unicellular or filamentous bacteria-like organisms. All are photosynthetic and some fix atmospheric nitrogen. Essential primary producers.

Kingdom Protista (*protist-*, first of all)

Eukaryotes include multinucleate heterotrophs such as the slime molds and water molds, unicellular heterotrophs such as the protozoa, and unicellular or multicellular autotrophs such as the algae. Reproduction is asexual and, in some species, sexual. This kingdom is often described as a "grab bag" of organisms since many evolutionary lines are included and few generalities apply. A true consensus does not yet exist on which groups of organisms should be placed within this kingdom. Approximately 60,000 living species and an additional 50,000 extinct species are classified within the kingdom Protista.

The first seven divisions are plant-like "algal protists"; the next two divisions are "fungal protists," and the final four are animal-like" protozoan protists."

Division: Chlorophyta (*chlor-*, green, + *-phyt-*, plant): "green algae"; a very large group of photosynthetic and mainly aquatic organisms that are either unicellular, colonial, coencytic, or multicellular. This group is believed to be linked to the evolution of more advanced land plants because of similarities in chlorophyll content (*a* and *b*), accessory pigments (carotenoids), cell wall composition (cellulose and hemicellulose), and food reserve (starch). Sometimes classified in the plant kingdom.

Class: Chlorophyceae (*chlor-*, green, + *-phyc-*, seaweed, + *-eae*, denoting a taxonomic class): "green algae"; unicellular, colonial, or multicellular algae primarily found in freshwater. Reproduction is primarily sexual with the formation of a dormant zygote that eventually undergoes meiosis to produce haploid cells.

Class: Charophyceae (*Chara*, a genus name meaning marsh reed, + *-phyc-*, seaweed, + *-eae*, denoting a taxonomic class): "stoneworts"; unicellualar or multicellular algae primarily found in freshwater. The body of some forms are often encrusted with calcium carbonate.

Class: Ulvophyceae (*Ulva*, a genus name meaning marsh plant, + *-phyc-*, seaweed, + *-eae*, denoting a taxonomic class): "sea lettuce"; coenocytic or multicellular marine algae. The body is often large and "leaf-like."

Division: Bacillariophyta (*Bacillario*, a genus name, + *-phyt-*, plant): "diatoms"; photosynthetic and unicellular with unique double shells heavily impregnated with silica, the two halves of which fit together like a Petri dish. Food is stored in the form of oil that also aids buoyancy. Essential aquatic primary producers.

Division: Chrysophyta (*Chryso-*, golden yellow, + *-phyt-*, plant): "golden-brown algae"; a diverse group of unicellular photosynthetic organisms, including flagellated, nonmotile, and amoeboid forms. Photosynthetic pigments include *a* and *c* and the accessory pigment fucoxanthin. Essential aquatic primary producers.

Division: Dinoflagellata (*dino-*, rotating, + *-flagell-*, whip, + *-ata*, pertaining to): "spinning algae"; photosynthetic and unicellular with a unique arrangement of two flagella, one of which beats in a groove that encircles the organism, causing a spinning action. Essential aquatic primary producers and a major component of phytoplankton. May also be referred to as Pyrrophyta.

Division: Euglenophyta (*eu-*, true, + *-glen-*, eyeball, + *-phyt-*, plant): "euglenoids"; photosynthetic (secondarily heterotrophic) and unicellular organisms with one or more light sensitive eyespots; highly motile. Fresh-water primary producers.

Division: Phaeophyta (*phae-*, dark, dusk, + *-phyt-*, plant): "brown algae"; photosynthetic multicellular marine organisms with a unique combination of pigments including chlorophyll *a* and *c* and the brown pigment fucoxanthin; food is stored as laminarian starch. Many large species exist with complex tissue differentiation; commonly referred to as kelp. Sometimes classified in the plant kingdom.

Division: Rhodophyta (*rhodo-*, red, + *-phyt-*, plant): "red algae"; photosynthetic marine organisms with a unique combination of pigments including chlorophyll *a* and *c* and the red pigment phycobilin; food is stored as floridean starch. Flagellated cells do not occur in any species. Most species are characterized by having fine, feathery branches. Sometimes classified in the plant kingdom.

Division: Myxomycota (*myx-*, slime, + *-myc-*, fungus): "acellular slime molds"; heterotrophic amoeboid organisms that form multinucleated plasmodia that move and ingest food particles by forming pseudopodia-like appendages. Sometimes classified in the Fungi kingdom because reproduction is by spores.

Division: Acrasiomycota (*acras-*, bad mixture, + *-myc-*, fungus): "cellular slime molds"; heterotrophic organisms that exist as individual amoeboid cells that swarm to form cellular plasmodia. The plasmodia move and ingest food particles by forming pseudopodia-like appendages. Sometimes classified in the Fungi kingdom because reproduction is by spores.

Division: Sacrodina (*sarcodin-*, like flesh): "amoeboids"; mainly marine unicellular heterotrophs that move and feed by forming pseudopodia; some species form hard perforated siliceous outer shells. Reproduction is both sexual and asexual.

Division: Mastigophora (*mastix*, whip, + *-phor-*, to bear): "flagellates"; unicellular heterotrophs that move by means of flagella; some are free living but most are parasitic including *Trypanosoma* (the cause of African sleeping sickness). Reproduction is both sexual and asexual.

Division: Ciliophora (*cili-*, a small hair, + *-phor-*, to bear): "ciliates"; unicellular heterotrophs that move by means of cilia. Reproduction is usually asexual but sexual conjugation is also common.

Division: Sporozoa (*spor-*, spore, + *-zo-*, animal): nonmotile unicellular parasites characterized by a dormant spore stage in the life cycle; includes *Plasmodium* (the cause of malaria).

Kingdom Fungi (*fung-*, mushroom, + *-us*, thing)

Eukaryotic multicellular (except for unicellular yeasts) organisms with vegetative bodies composed of a mass of multinucleated or haploid hyphal filaments; either saprophytic or parasitic heterotrophs with chitin present in the cell walls. Reproduction is sexual or asexual by spores produced in fruiting bodies. All fungi play an important role in decomposition. Over 100,000 species are classified within the kingdom Fungi.

Division: Chytridiomycota (*chytridi-*, little pot, + *-myc-*, fungus): "chytrid water molds"; multinucleated aquatic heterotrophes with a vegetative body usually composed of a small single-celled sporangium with many rhizoids. A true mycelium never forms. Reproduction is primarily asexual by zoospore formation and sexual by flagellated gametes.

Division: Oomycota (*oo-*, egg, + *-myc-*, fungus): "water molds"; aquatic or terrestrial heterotrophs with multinucleated (or diploid) filamentous bodies and cell walls composed of cellulose. Reproduction is by flagellated asexual spores and nonmotile gametes; the female gamete (oocyte) is characteristically large.

Division: Zygomycota (*zyg-*, yoke, + *-myc-*, fungus): "black bread molds"; terrestrial heterotrophs with multinucleated hyphae (septae form only during the development of reproductive structures); often found feeding off starchy material such as bread and potatoes. Some species form endomycorrhizal associations with plants.

Division: Ascomycota (*asc-*, cup, sac, + *-myc-*, fungus): "cup fungi"; primarily terrestrial organisms with multinucleated hyphae often fused together to form complex fruiting bodies called ascocarps. Sexual reproduction involves the formation of haploid spores within the ascocarp. Yeasts are the only unicellular acomycetes and reproduce asexually by budding. Molds, powdery mildews, morels, and truffles are classified within this division.

Division: Basidiomycota (*basid-*, club, + *-myc-*, fungus): "club fungi"; terrestrial organisms including mushrooms, toadstools, shelf fungi, rusts, and smuts. All reproduce sexually by forming basidia in which meiosis occurs or on which haploid spores are born.

Division: Deuteromycota (*deuter-*, second, + *-myc-*, fungus): "fungi imperfecti"; a collection of fungi in which sexual reproduction does not occur or has not been observed. It is has been suggested that many species may actually be members of the Ascomycota.

Lichen (not a true taxonomic grouping): a composite organism described as an integrated symbiosis between a fungal partner (usually an Ascomycete) and an algal partner (either a green algae or a Cyanobacteria). This symbiosis has often been considered mutualistic but recent evidence suggests the fungal partner may be parasitizing the algae.

Kingdom Plantae (*plant-*, a sprout)

Multicellular photosynthetic eukaryotes primarily adapted to living on land (terrestrial). All plants are characterized by similarities in chlorophyll content (*a* and *b*), accessory pigments (carotenoids), cell wall composition (cellulose and hemicellulose), and food reserve (starch). Advanced tissue and organ differentiation occurs in most species. Reproduction is primarily sexual with two phases in the life cycle; the gametophyte (gamete-producing) generation and the sporophyte (spore-producing) generation. Plants are essential as primary producers in all ecosystems; over 270,000 species are classified in the kingdom Plantae.

Division: Bryophyta (*bry-*, moss, + *-phyt-*, plant): "mosses, liverworts, and hornworts"; multicellular nonvascular plants with photosynthetic pigments and food reserve similar to green algae (see Chlorophyta). True roots, stems, and leaves are lacking. Distribution is restricted to very moist (almost semiaquatic) habitats because liquid water is needed to ensure fertilization (sperm are motile). The sporophyte is usually dependent on the gametophyte for nutrients.

Class: Hepaticopsida (*hepat-*, liver, + *-ic-*, pertaining to, + *-opsida*, denoting a taxonomic class): "liverworts"; the dominant gametophyte generation is usually flattened dorsiventrally and dichotomously branched; sporophytes are relatively simple in structure and short-lived. Reproduction is both asexual (fragmentation and gemmae) and sexual.

Class: Anthoceratopsida (*antho-*, flower, + *-cerat-*, horn, + *-opsida*, denoting a taxonomic class): "hornworts"; the dominant gametophyte generation is "leafy." Sporophytes originate from simple meristematic tissue, possess stomata and may be long-lived. Reproduction is asexual (fragmentation) and sexual.

Class: Muscopsida (*mus-*, moss, + *-opsida*, denoting a taxonomic class): "mosses"; the dominant gametophyte generation is "leafy" with long multicellular rhizoids on the ventral surface; complex vascular tissue is absent. The sporophyte has efficient spore dispersal mechanisms as well as stomata. Reproduction is asexual (fragmentation) and sexual. Members of this class are the most common and widespread of the kingdom Bryophyta.

Division: Psilophyta (*psil-*, naked, + *-phyt-*, plant): "whisk ferns"; considered to be the most primitive plants alive that possess vascular tissue (xylem and phloem). The dominant sporophyte generation is homosporous and leafless with dichotomously branched stems that continue below ground as rhizomes. The gametophyte generation is microscopic and independent of the sporophyte; sperm are motile and require liquid water to ensure fertilization.

Division: Lycophyta (*lyco-*, wolf, + *-phyt-*, plant): "club mosses" and "quill worts"; extremely diverse in appearance with both homosporous and heterosporous species. The dominant sporophyte generation in most species produces a specialized fertile "club-like" structure (stobilus) that produces spores. The gametophyte generation is microscopic and independent of the sporophyte; sperm are motile and require liquid water to ensure fertilization.

Division: Sphenophyta (*sphen-*, wedge, + *-phyt-* plant): "horsetails"; homosporous plants with scale-like nonphotosynthetic leaves and siliceous ribs running the length of the jointed stem. The gametophyte generation is microscopic and independent of the sporophyte; sperm are motile and require liquid water to ensure fertilization.

Division: Pterophyta (*pter-*, fern, + *-phyt-*, plant): "ferns"; mainly homosporous vascular plants with a conspicuous sporophyte generation possessing feathery fronds. The gametophyte generation possesses multicelluar gametangia and is very small and independent of the sporophyte; sperm are motile and require liquid water to ensure fertilization.

The next four plant divisions are often grouped within the obsolete and unofficial taxa of **Gymnospermae** (*gymn-*, naked, + *sperm-*, seed) because all members are flowerless plants that produce naked seeds (not enclosed within a fruit but rather on the surface of modified leaves on cones).

Division: Coniferophyta (*con-*, cone, + *-fer-*, to bear, + *-phyt-*, plant): "conifers"; seed plants with a dominant sporophyte generation and a highly reduced gametophyte generation (several cells housed within cones) dependent on the sporophyte until mature; abundant cambial activity produces very tall and sturdy sporophytes ("trees"). The leaves are highly reduced and "needle-like." Most species are monoecious with both male (pollen) and female (seed) cones produced on a single tree. Sperm are nonmotile and reach the female gamete (within an ovule) through a pollen tube.

Division: Cycadophyta (*cycas-*, palm-like +, *-phyt-*, plant): "seed ferns"; seed plants with a dominant sporophyte generation and a highly reduced gametophyte generation (several cells housed within cones) dependent on the sporophyte until mature. The leaves are large and "palm-like." All species are dioecious with separate male (pollen) and female (seed) cone-bearing plants. Sperm are motile and reach the female gamete (within an ovule) through a pollen tube.

Division: Ginkgophyta (from the Chinese, *yin*, silver, + *hing*, apricot, + *-phyt-*, plant): "maidenhair tree"; seed plants with a dominant sporophyte generation and a highly reduced gametophyte generation (several cells housed within cones) dependent on the sporophyte until mature. The leaves are large and fan-shaped with dichotomous venation. Sperm are motile and reach the female gamete (within an ovule) through a pollen tube.

Division: Gnetophyta (*gnetum*, a genus name, + *-phyt-*, plant): "gnetinas"; seed plants with a dominant sporophyte generation and a highly reduced gametophyte generation (several cells housed within cones) dependent on the sporophyte until mature. Several advanced angiosperm characteristics include nonmotile sperm and specialized xylem cells called vessels.

The final and most advanced plant division is often grouped within the obsolete and unofficial taxa of **Angiospermae** (angi-, enclosed, + *-sperm-*, a seed) because all members are flowering plants that produce seeds enclosed within a fruit.

Division: Anthophyte (*antho-*, flower, + *-phyt-*, plant): "flowering plants"; seed plants with a dominant sporophyte generation and a highly reduced gametophyte generation (several cells housed within flowers) dependent on the sporophyte until mature. Vegetatively an extremely diverse group of plants including woody and herbaceous forms; annuals, biennials, and perennials; herbs, shrubs, and trees. The process of double fertilization is unique to flowering plants and results in the formation of an embryo surrounded by endosperm tissue that nourishes the embryo.

Class: Monocotyledonae (*mon-*, one, + *-cotyle-*, cup): "monocots"; primarily grass-like plants with floral parts in multiples of three; parallel leaf venation, one cotyledon, and very little or no secondary growth (no woody species except for the palm family).

Class: Dicotyledonae (*di-*, two, + *-cotyle-*, cup): "dicots"; plants with floral parts in multiples of four or five; net-like leaf venation, two cotyledons, and woody species are very common.

Kingdom Animalia (*anim-*, breath, life, + *-al,* pertaining to)

Multicellular heterotrophic eukaryotes most of which are motile. All animals are characterized by having relatively flexible cells with no rigid cell wall present as in plants. Advanced tissue, organ, and organ system differentiation occurs in most species. Reproduction is primarily sexual, with diploid organisms producing haploid gametes (sperm and eggs) that unite to form a zygote. Over 1.5 million species are classified in the kingdom Animalia.

The first phylum (Porifera) is sometimes classified in the subkingdom **Parazoa** (*para-*, beside, + *-zo-*, animal) because they lack definite symmetry and organs. All the remaining phyla are sometimes classified in the subkingdom **Eumetazoa** (*eu-*, true, + *-meta-*, after, + *-zo-*, animal) because they all have radial or bilateral symmetry, and most have organs.

Phylum: Porifera (*por-*, channel, + *-fer-*, to bear): "sponges"; colonial or solitary aquatic organisms with rigid body walls supported by silica or calcium carbonate spicules. The body of the sessile adult is perforated with countless pores lined with flagellated cells (choanocytes) that cause food-carrying water to circulate through the organism. Reproduction is asexual and sexual.

Phylum: Cnidaria (*cnid-*, stinging nettle, + *-ar,* pertaining to): "jellyfish" and "corals"; radially symmetrical aquatic (virtually all marine) organisms with two distinct tissue layers; the only animals to have stinging cells (cnidocytes) that are used for defense and feeding. Most have two distinct body forms in the life cycle, a sessile polyp and a free living medusa adult. Reproduction is asexual and sexual.

Class: Hydrozoa (*hydr-*, water, + *-zo-*, animal): "hydroids," "Portuguese man-of-war"; organisms forming dense colonies of polyps (hydroids) possessing a regular alternation of asexual and sexual forms; the medusa stage may be absent or short-lived.

Class: Scyphozoa (*scyph-*, cup, + *-zo-*, animal): "jellyfish"; marine organisms with the free-living medusa stage dominant over the inconspicuous and simple polyp stage.

Class: Anthozoa (*antho-*, flower, + *-zo-*, animal): "corals" and "sea anemones"; organisms forming dense colonies of polyps (hydroids); medusa stage totally lacking.

Phylum: Ctenophora (*cten-*, comb, + *-phor-*, to bear): "sea walnuts" and "comb jellies"; radially symmetrical aquatic marine organisms with eight comb-like rows of cilia used for locomotion. All forms are translucent, gelatinous, and often bioluminescent.

Phylum: Platyhelminthes (*platy-*, flat, + *-helminth-*, worm): "flatworms"; bilaterally symmetrical, acoelomate organisms flattened dorsiventrally; ubiquitous in distribution (terrestrial, marine, freshwater, and parasitic). The simplest animals to have three germ layers and organs. Reproduction is asexual and sexual with most forms being hermaphroditic.

Class: Turbellaria (*turbell-*, stir, row, + *-ar,* pertaining to): "planarians"; aquatic free-living flatworms; ciliated, carnivorous. Reproduction is asexual by fission of the body, and sexual.

Class: Trematoda (*tremat-*, a hole, + *-odo,* swollen): "flukes"; parasitic flatworms with a well developed digestive tract; complex life cycles involving two or more hosts is common.

Class: Cestoda (*cest-*, girdle, + *-odo,* swollen): "tapeworms"; parasitic flatworms without a well developed digestive tract; food is absorbed through body surfaces; complex life cycles involving two or more hosts is common.

Phylum: Nematoda (*nemat-*, thread, + *-odo,* swollen): "roundworms"; bilaterally symmetrical, unsegmented cylindrical worms; ubiquitous in distribution (terrestrial, marine, freshwater, and many detrimental parasitic forms). The simplest animals to have a complete digestive tract (mouth and anus); pseudocoelomate. Reproduction is sexual.

Phylum: Loricifera (*loric-*, corselet, + *-fer-*, to bear): microscopic, bilaterally symmetrical animals living on the ocean floor. Discovered in 1982; characterized by a plate-covered body with numerous hair-like projections and a retractable tube-like mouth.

Phylum: Rotifera (*rot-*, wheel, + *-fer-*, to bear): "rotifers" or "wheel animals"; microscopic, bilaterally symmetrical worm-like animals with a "wheel" of cilia around the "mouth" region. Primarily freshwater in distribution but may also be terrestrial, marine, or symbiotic within some mosses.

Phylum: Molluska (*mollusc-*, a shellfish, soft): bilaterally symmetrical, unsegmented coelomate animals with a mantle and muscular foot, some with heads; widely distributed (marine, freshwater, and terrestrial). Many form shells, either external (clams) or internal (squid) that are used for support and protection. Reproduction is sexual.

Class: Polyplacophora (*poly-*, many, + *-plac-*, plate, + *-phor-*, to bear): "chitons"; marine animals with eight overlapping calcareous plates embedded in the dorsal surface of the mantle. The general body plan is relatively simple, elongated, with many gills. This class may also be referred to as **Amphineura** (*amphi-*, both, + *-neur-*, nerve).

Class: Bivalvia (*bi-*, two, + *-valv-*, folding doors): "bivalves" including clams, oysters, mussels, scallops; aquatic filter feeders that have two hinged shells, a highly reduced head, and a muscular foot. Generally sessile as adults but motile in the small larval stage. Reproduction is sexual.

Class: Gastropoda (*gastr-*, stomach, + *-pod-*, foot): "snails and slugs"; asymmetrical animals usually possessing a spiral shell and a well developed head with eyes and other sensory organs; widely distributed (marine, freshwater, and terrestrial) and generally herbaceous. Reproduction is sexual.

Class: Cephalopoda (*cephal-*, head, + *-pod-*, foot): "squid and octopuses"; highly developed marine predators with an enlarged head composed of a complex brain, eyes, mouth with two horny jaws, and eight (octopus) to ten (squid) arms and/or many tentacles (nautilus). The shell may be external (nautilus), internal (squid), or absent (octopus).

Phylum: Annelida (*annul-*, ring, + *-ide,* related to): "segmented worms"; bilaterally symmetrical, segmented coelomate animals with a head, complete digestive tract, and a closed circulatory system. The head is well developed with a brain (cerebral ganglia) and numerous sensory organs, including eyes in some species. Reproduction is sexual; asexual reproduction through fragmentation may occur in some forms.

Class: Polychaeta (*poly-*, many, + *-chaet-*, hair): "polychaetes" including sandworms and plumed worms; virtually all are marine forms with well developed heads with numerous sense organs including eyes. Lateral flap-like appendages called parapodia are associated with locomotion and respiration.

Class: Oligochaeta (*olig-*, few, + *-chaet-*, hair): "earthworms"; terrestrial, marine, and freshwater worms lacking a distinct head, external sense organs, and external appendages (parapodia). The body also has few bristle-like appendages (chaeta). Reproduction is sexual with most forms being hermaphroditic; asexual reproduction through fragmentation may also occur.

Class: Hirudinea (*hirud-*, leech, + *-ine-*, having the character of): "leeches"; worms flattened dorsiventrally with reduced segmentation; modes of feeding include predation, scavenging, and exoparasitism. One or more suckers are used for attaching to and feeding off of their host.

Phylum: Arthropoda (*arthr-*, joint, + *-pod-*, foot): "arthropods"; the largest phylum in the animal kingdom with over one million species identified, all of which are bilaterally symmetrical coelomates with paired jointed appendages, and segmented bodies covered with a chitinous exoskeleton. All organ systems are present with some highly developed. Ubiquitous in distribution (terrestrial, aerial, marine, freshwater, and parasitic). Reproduction is primarily sexual.

The first two classes are sometimes classified in the subphylum **Chelicerata** (*chel-*, claw, + *-cerat-*, horn) because all members have two main body parts and six pairs of appendages, including a pair of chelicerae (fang-like appendages used for biting). Due to the numbers and variation among the crustaceans, they are sometimes classified in their own subphylum **Crustacea.** The last three classes are sometimes classified in the subphylum **Uniramia** (*un-*, one, + *-ram-*, branch) because all members have single-branched appendages.

Class: Merostomata (*mer-*, part, + *-stom-*, mouth, + *-ate,* characterized by having): "horseshoe crabs"; a relatively small and primitive group of aquatic arthropods with five pairs of walking legs, one pair of chelicerae (fangs or pincers) used for feeding and defense, compound eyes, and book gills.

Class: Arachnida (*arachn-*, spider, + *-ida,* relating to): "spiders, ticks, mites, and scorpions"; primarily terrestrial air-breathing carnivores with four pairs of walking legs, one pair of chelicerae (fangs or pincers) used for feeding and defense, and one pair of pedipalps that are often sensory but may also be used for manipulating food.

Class: Crustacea (*crust-*, shell, + *-ac,* resembling): "crayfish, lobsters, shrimp, barnacles, and many others"; primarily aquatic arthropods with three distinct body regions: the head with numerous highly developed sensory organs such as compound eyes; the thorax with two or three pairs of appendages; and the abdomen with one pair of appendages or none. All appendages are biramous (two-branched).

Class: Chilopoda (*chil-*, lip, + *-pod-,* foot): "centipedes"; relatively aggressive predators with a distinct head that bears a pair of fangs and associated poison glands, and 15 to 177 body segments, each with a pair of jointed walking appendages.

Class: Diplopoda (*diplo-*, double, + *-pod-,* foot): "millipedes"; primarily herbaceous arthropods with a distinct head and a trunk composed of 20 to 200 segments, each with two pairs of jointed walking appendages.

Class: Insecta (*insect-*, cut into): "insects," including ants, bees, termites, grasshoppers, flies, etc.; by far the largest group of organisms (over 900,000 described species) with the greatest degree of diversity of form. All insects are terrestrial and have three distinct body regions: the head with numerous highly developed sensory organs such as paired compound eyes and antennae; the thorax with three pairs of appendages and often two pairs of wings; and the abdomen containing reproductive organs. Many species have complex life cycles involving metamorphosis.

Phylum: Echinodermata (*echin-*, spiny, + *-derm-*, skin, + *-ate*, characterized by having): "sea stars, sea urchins," and related forms; complex marine coelomate animals with penta-radially symmetrical adults and bilaterally symmetrical larvae. The unique water vascular system provides a means of locomotion by pressurizing numerous tube feet that extend through the perforated plates of the endoskeleton. Most forms have long movable spines for protection. Reproduction is primarily sexual although many forms are capable of regeneration.

Class: Crinoidea (*crin*, lily, + *-oid-*, resembling): "sea lilies and feather stars"; sessile cup-shaped filter feeders with 5 to more than 200 feathery appendages located around the margins of the cup. Most species are extinct.

Class: Asteroidea (*aster-*, star, + *-oid*, resembling): "sea stars or starfish"; star-shaped organisms with five (or multiples of five) appendages radiating out from a central disk. Rows of tube feet on the lower surface of each appendage provide a relatively slow method of locomotion.

Class: Ophiuroidea (*ophi-*, serpent, + *-ur-*, tail, + *-oid-*, resembling): "brittle stars"; star-shaped organisms with long, slender, and flexible appendages radiating out and sharply delineated from the central disk. The snake-like motion of the appendages provides a relatively rapid horizontal movement.

Class: Echinoidea (*echin-*, spiny, *-oid-*, resembling): "sea urchins and sand dollars"; spherical or round and flat organisms with a rigid external covering composed of many movable spines; long slender tube feet extending through perforations in the external covering provide for a slow but efficient method of locomotion.

Class: Holothuroidea (*Holothuria*, a genus name meaning a kind of plant-like animal, + *-oid-*, resembling): "sea cucumbers"; long cylindrical soft-bodied organisms lacking the tube feet and spines characteristic of most echinoderms.

Phylum: Chaetognatha (*chaet-*, hair, + *-gnath-*, jaw): "arrow worms"; small, translucent, free-living planktonic marine worms with several advanced characteristics including bilateral symmetry, a true coelom, complete digestive tract, and a mouth with powerful jaws. Reproduction is sexual.

Phylum: Hemichordata (*hemi-*, half, + *-chord-*, cord, + *-ate*, characterized by having): "acorn worms"; soft-bodied marine worms divided into three distinct body regions (proboscis, collar, and trunk) and numerous advanced characteristics including bilateral symmetry, a true coelom, complete digestive tract, ventral and dorsal nerve cords, a pharynx, and gill slits. Reproduction is sexual.

Phylum: Chordata (*chord-*, cord, + *-ate*, characterized by having): "chordates"; bilaterally symmetrical coelomate animals having at some stage in their development pharyngeal gill slits (or pouches), a notochord, a hollow dorsal nerve cord, and a tail. Reproduction is sexual.

Subphylum: Cephalochordata (*cephal-*, head, + *-chord-*, cord, + *-ate*, characterized by having): "lancelets" (amphioxus); filter feeding, fish-like marine organisms with a notochord and a nerve cord running the entire length of the body, a pharynx with gill slits, and segmented body muscles.

Subphylum: Urochordata (*ur-*, tail, + *-chord-*, cord, + *-ate*, characterized by having): "tunicates and sea squirts"; sessile, filter-feeding, sac-like marine organisms (the bilaterally symmetrical larvae resemble free-swimming tadpoles and have a distinct nerve cord and notochord).

Subphylum: Vertebrata (*vertebr-*, a joint, + *-ate*, characterized by having): "vertebrates"; the best known group of animals, the adults of which have a jointed vertebral column made of bone or cartilage that replaces the notochord, and a cranium (skull) surrounding the well-developed brain.

Class: Agnatha (*a-*, without, + *-gnath-*, jaw): "lampreys and hagfish"; aquatic, eel-like fishes lacking limbs, scales, and jaws; all have cartilaginous skeletons. Parasites on other fishes, or scavengers.

Class: Chondrichthyes (*chondr-*, cartilage, + *-ichthy-*, fish): "sharks, skates, and rays"; almost all marine fishes with cartilaginous skeletons, complex copulatory organs, internal fertilization, but lacking air bladders; efficient predators with well-developed jaws.

Class: Osteichthyes (*oste-*, bone, + *-ichthy-*, fish): "bony fishes"; jawed fishes with skeletons composed of bone, all forms possess air bladders for buoyancy. Abundant in freshwater as well as marine habitats.

Class: Amphibia (*amphi-*, both, + *-bi-*, life): "amphibians," including frogs, toads, and salamanders; ectothermic jawed tetrapods that respire by gills in the larval stage and by lungs (and moist skin) in the adult stage. Fertilization is external and the aquatic larvae undergo a complete metamorphosis to become terrestrial adults.

Class: Reptilia (*reptil-*, to crawl): "reptiles," including snakes, turtles, lizards, and alligators; ectothermic jawed tetrapods (except for snakes and some lizards) that lay amniotic eggs and respire through lungs. All are covered with scales; most are terrestrial.

Class: Aves (*avis*, bird): "birds;" endothermic tetrapods with forelimbs modified into wings and skin covered with feathers. All lay amniotic eggs.

Class: Mammalia (*mamm-*, breast, + *-al*, pertaining to): "mammals;" endothermic tetrapods (various limb modifications include wings, arms, and fins) that nourish their young with milk secreted by the mother. The majority of mammals are placental and nourish their developing young in the womb through the placenta. All have hair covering their skin and are considered relatively complex both functionally and behaviorally.

Further References

There are literally hundreds of word reference books for the life sciences, many of which are specific to a particular discipline or subject area. The following is a list of general references you may find useful.

1. Ayers, D. M. 1985. *Bioscientific Terminology: Words From Latin and Greek Stems.* The University of Arizona Press. Tucson, Arizona.
2. Borror, J. 1971. *Dictionary of Word Roots and Combining Forms.* Mayfield Publishing Co. Palo Alto, California.
3. Holmes, S. 1991. *Henderson's Dictionary of Biological Terms.* 10th ed. Van Nostrand Reinhold Co. New York, New York.
4. Jaeger, E. C. 1978. *A Source-Book of Biological Names and Terms.* 3d ed. Charles C. Thomas. Springfield, Illinois.
5. Squires, B. P. 1986. *Basic Terms of Anatomy and Physiology.* 2d ed. W. B. Saunders. Toronto, Ontario.
6. Thomas, C. L. (Editor). 1989. *Taber's Cyclopedic Medical Dictionary.* 19th ed. F. A. Davis Co. Philadelphia, Pennsylvania.
7. Walker, P. M. B. (Editor). 1989. *Chamber's Biology Dictionary.* Chambers & Cambridge. Cambridge, England.